# *the secret of consciousness*

**HOW THE BRAIN TELLS 'THE STORY OF ME'**



*paul ableman*

MARION BOYARS
LONDON • NEW YORK

Published in Great Britain and the United States
in 1999 by Marion Boyars Publishers
24 Lacy Road, London SW15 1NL
237 East 39th Street, New York NY 10016

Distributed in Australia and New Zealand by
Peribo Pty Ltd, 58 Beaumont Road, Mount Kuring-gai, NSW



British Library Cataloguing in Publication Data
Ableman, Paul
The secret of consciousness: how the brain tells 'the story of me'
1. Consciousness 2. Self 3. Memory
4. Psycholinguistics
I. Title
153

Library of Congress Cataloging-in-Publication Data
Ableman, Paul.
The secret of consciousness: how the brain tells 'the story of me / Paul Ableman.
p. cm.
1. Consciousness. 2. Memory. 3. Human information processing. 4. Psycholinguistics. I. Title.
BF311.A195 1999
153--dc21 98-54697
CIP

ISBN 0-7145-3053-0 Paperback (alk. paper)



Typeset in 11½/13½ pt Caslon by
Ann Buchan (Typesetters), Shepperton
Printed in Great Britain by
WBC Book Manufacturers, Bridgend, Mid Glamorgan

*t... of consciousness*

**By the same author**

I Hear Voices
Tornado Pratt
Twilight of the Vilp
The Doomed Rebellion
Vac

## Contents


**PART ONE**

Two 'I's' 9

The Mystery of Dreams 12

Thoughts That Come From Nowhere 19

The Ceaseless Work of the Brain 35

Night and Day 51

The Mind of Man 69

The 'Story of Me' 89

Beings 108

**PART TWO**

Malfunctions 125

Towards a Philosophical Context 145

Brief Bibliography 158

Index 159

# PART ONE

# Two 'I's'

This book is the long-delayed result of my having noticed a curious anomaly in a dream many years ago. I call it an 'anomaly' because that is how it seemed to me at the time. But now I know that it was much more than that. It was really a key to the truth about consciousness and about the human mind itself.

It is several decades since I awoke one morning and recalled that, in my dream, I had been able to see myself from a point of view outside my own body. I do not mean that I experienced what is called an 'out-of-body' experience. No, what occurred was that I found myself able to see my own image within the dream itself. Having once perceived that this rather uncanny perspective is possible, as I daresay other people have also done, I later found that this kind of dream occurred quite often.

I will give an example of such a dream to make my meaning absolutely clear. I might wake up and remember that I had just dreamed that I was eating a meal in the company of several other people. Nothing very remarkable about that. Yes, but in the dream I had been able to see myself at the table along with the other diners. I had somehow managed, in the dream, to be both 'myself' and the observer of myself. Within the structure of the dream I had become two people.

In waking life it is impossible (other than by using mirrors or other optical devices) to see oneself from outside the body. I had noted that in the dreams which contained 'two I's' I was not seeing a dream mirror image since I was sometimes able to see myself from behind and mirrors (at least single mirrors) cannot supply such a perspective. I had always assumed, as I think most people do, that the 'I' who has strange adventures in dreams remains essentially the same 'I' as the one who lives in the real world. Now this supposition seemed questionable.

Gradually, and over a period of years, the recurrence of the out-of-body point of view acted on my own curiosity and one day I put to myself a crucial question: if some of my dreams seem to involve two quite separate and distinct 'I's' — one of whom is the observer of the other — then which of them is the 'real I'? This question immediately suggested another which had rather an eerie tone to it: assuming that one of the two 'I's' is the 'real I' then who exactly is the other?

About a year ago, brooding about these 'self-visualizing' dreams, I decided to look into the matter. At that stage, it still did not occur to me that I would one day write a book which would take such a trifling puzzle, as I still thought of it, for its starting point. Certainly I had not the slightest inkling that such an idle quest, undertaken chiefly out of curiosity, would lead me to an insight into not only the origin of dreams but also the nature of the brain and the truly astounding way in which it generates the mind.

I am a professional novelist and dramatist but I have been keenly interested in science and technology all my life. I started to read the literature on dreams and dreaming to see if I could find any explanation for the 'two I's' phenomenon. In fact, I found little that was relevant but before long realized that there were not just two but four possible answers to the question: which is the real 'I' in 'two-I' dreams?

These were that the real 'I' could be the observer 'I' (not

actually 'visible' in the dream). Alternatively, he could be the observed 'I'. But he might also, split into two halves by some mechanism possibly connected with the fact that there are two hemispheres in the human cerebral cortex, be both the observed 'I' and the observing 'I'. Finally, the true 'I' might be neither observed nor observer.

It is now a year since I formulated the question to myself in this way. I have discovered, amongst much else, that the correct answer is that neither of the two 'I's that figure in some dreams is the 'real' one. This was, for me, the least likely answer and so I was astonished and even dismayed to learn that it was the right one.

However, my researches have shown me that the 'real I' — that is, the waking, autonomous human being — plays only a very small part in dreaming. The images of the self that occur in dreams have only a little more relationship to the true identity of the dreamer than any other images in dreams do. The rest of the first, and main, section of this book will disclose how I came to understand the reasons for this strange fact and also the disturbing, and even frightening, implications that flowed from the discovery.

# The Mystery of Dreams

It is a surprising fact that even at this late stage in the evolution of science, hardly anything is known with reasonable certainty about either sleep or dreaming. Part of the reason is doubtless the fact that for thousands of years the chief commentators on dreams have been not genuine philosophers and scientists but rather magicians, astrologers and plain charlatans.

There is, however, one group of reputable, if not in the strictest sense science-motivated, workers who have taken dreams very seriously. These are psychoanalysts and other psychotherapists. Since they work with what are regarded as disordered minds, and since dreams seem to provide a window into the mind, it has been virtually impossible for them to ignore dreams. But a sense of frustration can be detected in their comments on the subject. For the fact is that dreams do not perform according to reasonable expectations as regards providing useful insights into mental disturbances.

Peering through the window that dreams might be expected to open into the mind, what psychotherapists actually perceive are not clear signs of malfunction in otherwise orderly machinery but strange shifting phantasmagoria. From recorded, as well as experienced dreams, it has proved difficult if not impossible for psychotherapists to derive consistent diagnostic data.

The content of dreams is surreal and their message, if one can be ascertained, is invariably obscure and ambiguous. These qualities, which are precisely the ones that appeal to astrologers and other fringe religious practitioners (if also, more positively, poets, artists and others who work with the imagination) have consistently frustrated the attempts of reputable psychotherapists to use dreams to much advantage for therapeutic or diagnostic purposes.

However, notwithstanding the ambivalent attitude of many psychotherapists to dreams, a good deal of painstaking technical research into the nature of sleep has actually been undertaken. Ever since the thirties, electroencephalography has been used to record the patterns of electrical activity in the brains of sleepers. And this type of research has resulted in one discovery that is both fundamental and very important. Electroencephalograms have demonstrated that sleep generates regular cycles of measureable electrical activity in everyone's brain. One of these cycles is associated with what is known as REM (Rapid Eye Movement) sleep and it is during this type of sleep that dreams most often occur.

But this charting of electrical cerebral activity represents more or less all that science has thus far discovered about sleep and dreaming. As regards the deeper nature, never mind any possible function, of either sleep or dreaming, little advance in understanding has occurred since the dawn of history.

Basically sleep and dreaming remain to modern science as mysterious as they seemed to pagan priests and shamans. Moreover, the inherently puzzling nature of dreams is compounded by the fact that science has never managed to find a clear and demonstrable reason why human beings need sleep at all. In fact, sleep is not, as most people assume, necessary in order to provide bodily rest. The body can achieve adequate and satisfactory rest by simple relaxation unaccompanied by sleep.

Why then should a more or less profound unconsciousness, during which dreams may, but do not always, occur, take possession of every human being for a third of his or her life? Science simply has no generally accepted answer to, or even very plausible theory about, the matter. The basic truth is that dreams are rather despairingly considered by many researchers to be merely bizarre and inherently meaningless fragments of unrelated imagery, having little or nothing to do with any functionally important property of the brain.

The following are just a few examples of comments on the nature of dreams made by some of the leading modern thinkers concerned with the brain and its activity. Daniel Dennett, in his exhaustive work *Consciousness Explained* rather surprisingly simply ignores dreams. Douglas Hofstadter in his classic, and entertaining, modernist book *Gödel, Escher, Bach: An Eternal Golden Braid* suggests that 'Dreams are perhaps just . . . random meanderings about the ASU's of our minds.' An 'ASU', Hofstadter explains, is an imaginary map of the USA. (The equivalent for a sleeping Frenchman would presumably be an ECNARF?)

Isaac Asimov in his excellent and exhaustive *New Guide to Science* offers the tentative, but as I will try to show, partially correct suggestion that 'dreaming is a device whereby the brain runs over the events of the day to remove the trivial and repetitious that might otherwise clutter it and reduce its efficiency.' Perhaps the major modern thinker to concern himself with the question of the function of the brain and its relationship with consciousness, Roger Penrose, does not even list 'sleep' or 'dreaming' in the index of his most recent and mathematically challenging work *Shadows of the Mind*.

Jacob Empson, however, in his excellent textbook *Sleep and Dreaming*, exhaustively examines all theories, ancient and modern, concerning the nature of dreams but, although he has personally run the dream laboratory in Hull for many years, he does not offer any firm hypothesis as to the essential nature of dreams. The index to his book lists various possibilities of dream genesis such as 'day's residues' (in this he echoes Asimov), 'hunger and thirst', 'incorporation of stimuli', 'pre-sleep stimulation' and 'wish fulfilment'.

And yet it is manifestly the case that dreams are, in some way or another, generated by the brain and the brain is recognized by both scientists and laymen to be the most important organ that human beings possess. As regards size, energy consumption and the range and significance of its activities, the human brain is the organ that most clearly puts us in a special class amongst animals.

We know for certain that all the other organs that man possesses, with the possible exception of the pineal gland, are highly functional. They perform a specific job. They are, moreover, of a size and capacity precisely tailored to perform that job. Science has, by now, very largely succeeded in relating the structure of all the organs other than the brain to the quantity and nature of the work that they perform.

The heart, for example, is a pump designed — one might almost say engineered — to circulate an exactly designated quantity of blood so many times in any given period. It is not too large or too small for this job. It has no spare capacity or unneccessary structural elements. The same, with the appropriate changes, can be said of the liver, pancreas, kidneys etc. They have been designed, or more accurately evolved, to perform a specific job and they do it efficiently from birth until death.

Most of these organs are largely automatic in their cycle of activity. You do not have to tell your heart to open the right auricle and close the right ventricle. In fact, it is almost impossible to affect consciously the functioning of the heart. This is true of the liver and kidneys as well — although it is possible for the conscious mind to take over briefly control of the lungs. I hope to be able to demonstrate that brain activity shares with that of these other organs the quality of being largely under automatic control.

At present the prevailing scientific view of the brain is that it is exceedingly difficult, if not impossible, to relate its amorphous physical structure to the work that it performs. Moreover, the actual nature of that work remains largely obscure. Almost all who undertake the study of either brain function or brain structure seem to acquire before very long an uneasy sense that the brain is actually engaged in performing prodigies of work but that exactly what these prodigies really are is almost impossible even to guess at.

In fact, as I hope soon to demonstrate, it really is possible to determine the nature of the work performed by the human brain, to show how this work varies by night and by day and to specify precisely what activities the brain is engaged upon during every phase of its astonishing twenty-four-hour cycle. The key to all these discoveries, as we shall see, is the nature of dreams.

The brain is essentially a vast agglomeration of elaborately interconnnected cells called neurons. It is not possible, as it is with every other organ in the body, to obtain visual clues as to its function. The heart can be seen to be a pump and the kidneys a filter. To a greater or lesser extent this correlation between function and structure applies to all parts of the body except the brain.

This master organ remains functionally indeterminate. Even those regions of the brain which have been anatomically classified, such as the Corpus calosum, Pons varioli, the cerebrum and cerebellum themselves, have only very imprecise, and even arbitrary, borders. And the relationship between them and the kind of activity that goes on in them is always to some extent insecure.

It is well known that loss of function in one region of the brain may subsequently be duplicated in quite a different area. It is also a mysterious fact that a large region of the brain may be damaged with little apparent effect on behaviour while, in other cases, tiny lesions can produce devastating results. Metaphorically the brain can be thought of as rather like a city. Corresponding to the city's 'districts' ('theatre' 'shopping' 'residential' etc) the brain has regions given over to things like vision, speech, balance and many more. But the brain's districts, like the city's, seem to be only 'traditionally' located in a given region. They may in time be transferred to another part if some pressing need for relocation occurs.

As regards the nature of brain function, it is generally accepted that the brain processes data that it receives from the sense organs: eyes, ears, nose etc. Fairly obviously it also 'runs' the body, monitoring its activity and state of health and regulating its automatic and semi-automatic activities. But it is clear as well that the brain does vastly more than these relatively routine jobs.

This much can be inferred from the fact that the brain is the organ that generates all the attributes that distinguish a human being from other animals. These are things like mind, self-consciousness, personality, use of language and so on. It is, of course, difficult to discuss any of these concepts other than in abstract terms and it is virtually impossible to relate them either to specific areas of the brain or to the little we know about brain activity.

What, for example, other than its routine task of monitoring and regulating endogenous bodily systems, does the brain do during the third of our lives when we are asleep? Can its task during sleep (remembering that sleep takes up some twenty or thirty years of our life) really be no more than the manufacture of the enchanting and sometimes terrifying (when they are nightmares) stretches of surreal imagery we know as dreams? Or might it be that dreams are really a purposeful and possibly necessary product of the sleeping brain in the same sense that blood is the product of the bone marrow or insulin of certain pancreatic glands?

My starting point in this inquiry has been the belief that either a great deal more must be going on in the brain during sleep than the production of dreams or else that dreams must be of crucial importance to our lives in ways that have not as yet even been suspected. When I finally grasped the nature of the diurnal/nocturnal cycle of brain activity I realized that it involves both these alternatives.

In the first place, vastly more important activities than the production of dreams really are going on in the human brain during sleep. These activities are, in fact, so impressive that the first glimpse that I obtained of their scope and nature struck me as little short of awe-inspiring. But the dreams themselves, I ultimately came to realize, are essentially by-products of the functional task of the sleeping brain and have no direct importance. Nevertheless their fortuitous role in helping to mould human psychology, and beyond that human culture, has been of immense historical significance.

# Thoughts That Come From Nowhere

I began my search for the true nature of dreams by trying to find out in some detail how the waking brain operates. I realized that I was in possession of a piece of information that might serve as a potential starting point for this quest. This was an observation on my own waking mental functioning which I had first made many years before. It was, in fact, a perception analogous to the one about dreams which I described at the beginning of this book, that is the possibility of seeing oneself in a dream from a point of view that is outside the body. I classified this waking phenomenon rather unscientifically as 'thoughts that come from nowhere'.

I had been aware, possibly since early childhood, that at any moment in the waking day a thought that seemed irrelevant to its context might suddenly appear in my mind. This event is a common one and most readers are likely to have experienced it for themselves. If they have not personally noticed its occurrence then they will almost certainly have heard someone ask a rhetorical question such as: 'Now, why did I think of that just then?' or 'I wonder what made that thought come into my mind?' or something along these lines.

For a long time, as with the observation of out-of-body points of view in dreams, I accepted these apparently vagrant mental intrusions without paying them much attention. But then on one occasion an image appeared in my mind that did give me pause. There was nothing very startling about the image itself. Indeed, its very ordinariness was in part responsible for my surprise at perceiving it. It was a memory from childhood and it is now a long time since I originally experienced it. For this reason, I cannot be certain exactly what the image was or the context in which it occurred, but I do remember that it was the image of a flower or plant.

Perhaps it was the memory of bluebells in a wood that I had seen as an infant or of gingerly touching bright yellow gorse on a cliff top or of marvelling at the velvet smoothness and rich green of moss on an old stone wall. The only thing I am certain about is that I abruptly thought of a plant and that the initial visual or tactile experience with that plant had occurred in early childhood.

I asked myself if the memory was relevant to what I had been doing or thinking about at the exact moment at which it had occurred. But I failed to come up with any very clear link. My chief response to the incident was sheer amazement at the realization that an instant — as it were a molecule — of time, retaining all its original colour and poignancy, had lain dormant in my brain for more than thirty years, waiting, presumably, for an occasion when its reanimation might have been of some relevance (although I could not fathom what that relevance might be) to the events in my life at the time it recurred.

And then an even more momentous consideration occurred to me. How had my 'subconscious mind' — as I then still thought of it — managed to select, almost instantaneously and presumably appropriately, from a stock of what must certainly have consisted of millions, and conceivably of thousands of millions, of stored impressions this particular one? I was dimly aware that merely to ask the question was to achieve a perspective which suggested that the human brain, even in the context of our current experience of electronic computers, must possess virtually unimaginable powers of information processing and retrieval.

From that time onwards, I made something of a hobby of first noting, and then attempting to ascertain the provenance and the purpose of, such 'thoughts that come from nowhere'. I did not make a systematic study of the phenomenon, which I soon found was a regular part of my mental functioning, and it certainly never occurred to me that I would one day undertake an inquiry in which such thoughts would prove to be of central importance. But I did quite often ask myself why a particular impression, like the original one of a plant seen and appreciated in childhood, had suddenly become conscious. And I soon began to perceive that it was quite often possible to do so. No matter how subtle or marginal the thought or memory might seem on first inspection, closer examination would often connect it to one's life situation and hence to one's data needs at the time of its manifestation.

I soon realized, for example, that a great many thoughts coming from 'nowhere' concerned topography. I would often abruptly recall something as fleeting as a glimpse of a few yards of road that I had once, and perhaps only once, driven along in a car, possibly at high speed and perhaps many years in the past. At other times I would remember some landmark or conspicuous feature of a landscape. And I would, with increasing frequency as I began to examine more intensively the circumstances in which these memories occurred, be able to perceive that their recollection could in theory, if not always in practice, have been considered useful to me in some way.

For example, the memory of a stretch of road might occur while I was driving down another road which had common features with the original one or possibly when I was making another journey along the same road. Then again I might, on a country walk, be in need of a landmark to orientate myself and the memory of the one that came to mind would thus be potentially useful. I gradually reached the conclusion that what my 'subconscious mind' was continuously offering me was potentially useful clues towards the solutions to problems of location or direction that I either was, or might seem to be, encountering.

These clues would often prove to be of only slight, if any, use to me but they were always at least relevant. After a moment or two of reflection, I would usually be able to perceive why they had appeared in my conscious mind. In this way I began to realize that my brain must be perpetually analyzing my current situation in space and time and responsively presenting me with data that might help me in managing that situation as successfully as possible.

But why, I asked myself, were topography and travel so conspicuous in my recollections? Moving across the Earth's surface, and matters arising from such motion, naturally only occupied quite a small part of my time. I finally decided that the disproportion between the time I devoted to travel and the number of recollections it inspired might be the consequence of an atavistic tendency of the human brain. This would derive from the period when our remote forebears were nomadic inhabitants of the African savannah.

In such primitive conditions, it would have been vitally important to have a knowledge of local topography and to recognize landmarks. Such data would have been essential for maximizing safety in everyday life. It would have enabled the individual both to avoid dangers such as the presence of swamps or the lairs of predators and, positively, to find useful sites such as those where food, water or shelter could be obtained. According to this hypothesis, the reason why my modern brain persisted in offering me a stream of mostly only very marginally useful information about road junctions and landmarks, and quite often when I was not even engaged in travelling, was because my proto-human ancestors had required a continuous feed of cognate information for sheer survival.

But by no means all the 'thoughts that came from nowhere' concerned topography and nor were all of them first-time recollections of an original experience. I realized that many of them were, in fact, drawn from a regularly used repertory of recollected images which recurred in my mind whenever similar circumstances arose in my life. Much later in my investigations, but largely based on this perception, I realized that not all the impressions housed in what I was by then calling my 'archival memory' are equally prominent or accessible.

They seem, in fact, to be stored in terms of a hierarchy of usefulness and probably each time one is summoned for guidance it is, by some technique of emphasis — and most probably a chemical one — available to the brain, given greater definition and facility of access. Conversely an impression that is rarely or never summoned will probably dwindle in ready availability although not, as my original experience with the plant suggested, in clarity and vividness.

I soon discovered as I continued to monitor my thoughts that many of the 'intruder thoughts' were relevant to social or professional life, family life, literary or cultural life — indeed any and all aspects of experience. And I found that, if I made the attempt, I could usually, by concentrating hard (although even then sometimes only with considerable difficulty and perhaps only very tentatively), assign such recollections a functional role in what I was doing or thinking about at the time they occurred.

There remained, however, a small residue of 'thoughts which came from nowhere' for which, try as I would, I could find no rationale. They really did seem to have appeared quite gratuitously in my mind. And it was brooding about these 'rogue' interlopers which slowly led me towards an understanding of the extraordinary structure of the human brain and beyond this of the nature of the human mind. It still did not occur to me, however, that I was picking my way along a path that might ultimately lead me to an understanding of one of the oldest and most tormenting of philosophical and scientific questions: the relationship between the physical brain and the apparently immaterial mind.

My procedure in the matter of determining the reason for the appearance of an inexplicable 'thought that came from nowhere' was as follows: having examined such a rogue thought for some time and tried to the limit of my ability to credit it with some degree of associational relevance, I might then ultimately be compelled to decide that it simply lacked any. I would therefore have no alternative but to ask myself if in fact it had been deliberately offered to me by my subconscious or — but what was the alternative? I could only think of one: the possibility that the rogue had appeared in my consciousness as the result of simple error.

I found this somewhat distressing. Why? I ultimately decided it was because the brain and its powers are so central to the successful conduct of one's life that it is unsettling to feel that this master organ is fallible. At the most fundamental level, my dismay perhaps stemmed from the ancient and terrible fear of brain disorder — of madness — and this made me shy away from the interpretation of plain error. But, of course, common sense insisted that the brain, being a biological organ, must, like all other biological organs, be capable of malfunction. Ultimately therefore the possibility of simple error had to be considered.

Accepting the possibility of error in mental functioning turned out to be a very important staging post in my journey of discovery. For it led me to ask a crucial question that, when I attempted to find an answer to it, took me a long way further along the road towards understanding the brain's mode of operation.

I had already observed that a lot of the thoughts that, unsummoned, appeared in my mind could be discovered, by intense concentration, to have some degree of relevance to my psycho-historical situation at the moment of their appearance. Other such thoughts — albeit a small minority — apparently had little or no relevance and might therefore have reached my consciousness through simple error. So the question that now arose can be stated as follows: just how capacious was the total flow of thoughts coming to me from what, in deference to my growing perception of both its magnitude and its nature, I had begun to think of as my 'archival memory' rather than, in the closest conventional phrase, the unconscious mind.

I devoted a considerable amount of thought to this question and I slowly began to realize that the data stream entering consciousness from archival memory must, at the very least, be far greater than could be ascertained by simply monitoring one's own consciousness. This conclusion stemmed from the realization that most of the items of information manifesting themselves in consciousness were clearly not 'rogues'. This was because the rogues, as a result of their absolute or relative incongruity, are always conspicuous and hence quite easily detectable.

However, the majority of the thoughts that appeared even without having been deliberately summoned by the conscious mind were manifestly purposeful, that is they were relevant to mental processes that were taking place in consciousness at that very moment. And it was this very relevance which meant that a large part of the data flow reaching consciousness from archival memory might very easily be undetectable or at least very difficult to detect.

In other words, the relevant part of this incoming stream of thoughts would blend easily into the dominant pattern of thoughts in consciousness at any given moment and would thus be highly resistant to self-monitoring. This meant that by far the greater number of the 'thoughts that came from nowhere' would be camouflaged by their very relevance and might thus, under normal circumstances, not be noticed at all. The incidence of the thoughts that self-monitoring was capable of discerning would thus be indecisive for estimating the magnitude of the entire data stream issuing from archival memory and discharging into consciousness.

At this stage in my analysis of the process of consciousness-generation, a new question arose: did even the 'normal' data stream, whether composed of 'rogue' or relevant thoughts, reaching consciousness represent the entire data flow leaving the archival memory? Or might it be possible that more, and perhaps vastly more, data was summoned initially from archival memory than ever ultimately reached consciousness?

I knew, for example, that I could consciously summon up the memory of dozens of experiences of trees and plants which had stirred me deeply in childhood. It was therefore a reasonable assumption that my archival memory must hold memory traces of a great many such plant experiences having at least the same degree of intensity as the memory of the bluebells, gorse or moss which had actually come to me. For this reason it seemed virtually certain that some powerful selection process had operated to single out from the large, and probably vast, store of plant memories in my archival memory the specific one of 'bluebells' that had actually occurred to me on that occasion many years ago.

Unhappily it was too late for me even to guess at the circumstances which had dictated that the memory of the bluebells might have been considered potentially of some use to me when it had occurred. But it was still highly likely that the circumstances which had called up from my archival memory the image of the 'bluebells' on that occasion might have found many other stored memories concerning plants of similar relevance. And this in turn made it a reasonable hypothesis that although the bluebells originally had made an impact on my infantile consciousness that endowed it with some kind of emphasis, which was probably chemical in nature, in my memory archive, similar, if slightly less intense, qualities would have been shared by many other plant memories.

In other words, the manifest ease of retrievability of 'the bluebells' would almost certainly have applied to many similar memories. If the bluebells had been specially relevant, then other thronging memories of childhood experiences of plants must have been very nearly, if perhaps not quite, as relevant. And yet only the bluebells had reached my consciousness. It thus seemed very likely that the total quantity of cross-referenced information appropriate to any life situation, and hence probably activated from the memory archive by the exigencies of each passing moment of experience, would certainly be greater, and quite possibly immensely greater, than the amount of information that ever actually 'got through' into consciousness.

But 'got through' what exactly? Clearly a selection process could be inferred since it would be inconceivable for all the data relevant to each experiential instant that was held in each archival memory not only to be mobilized on a non-stop basis but actually to reach consciousness. If this were the case then the mind would simply be swamped by torrents of unmanipulable information. A brief glimpse of a bluebell wood might have usefully complemented a much later experience. Crowding memories of all the stored 'gardens of the mind' would simply have generated chaos. Moreoever, could one really believe that a veritable Niagara of data was pouring moment by moment into each consciousness without that mind even noticing the fact? This seemed to me inconceivable.

I therefore had to conclude that every item of data summoned from archival memory first has to 'get through' some kind of screening mechanism which is continually at work filtering the input and allowing only the most urgent and immediately relevant material ultimate access to consciousness. Such a hypothesis might, at the cerebroneurological level, help explain the immensely sophisticated and elaborate method by which nerve signals are transmitted through neural pathways.

This transmission involves the use of a large variety of different chemical neurotransmitters to bridge synapses between nerve cells. In addition, the signal is advanced by means of the creation and then neutralization of electrical potential as a result of sodium and potassium ions being pumped into and out of carrier paths. The well-nigh stupefying complexity of this system might, however, be explicable as a response to the permanent need to edge memories that are chemically coded for the intensity and quality of the original experience closer and closer to consciousness without necessarily ever quite getting them there.

It also seemed likely that, since the 'thoughts that come from nowhere' are clearly not specifically and individually summoned by the mind (as when a conscious effort is made to remember some forgotten datum), this entire data flow issuing from the archival memory must be evoked automatically. The most plausible technique for selecting and despatching torrents of data at the speed implied by the functioning of the entire processs would be by the use of some kind of transponder mechanism.

To achieve this, all sensory data entering the cerebroneural system would have to be diverted through twin pathways, one to consciousness and the second to the archival memory itself (actually a third pathway is involved but we can defer consideration of this to a later stage). The second input, the one impacting on the archival memory, would then immediately trigger the despatch of complementary guidance data from the archive.

Thus an input from sensory monitoring of the environment which revealed, say, the start of a shower of rain would automatically mobilize a response from all the data cells in archival memory which contained any meteorological or other relevant data. This would include information concerned with the individual's previous experience of inclement weather, local topography to be scanned for possible shelter, data as to conflicting demands for the person's presence which might make it advisable to ignore the bad weather and so on.

It is easy to see that a 'galaxy' of information of this kind, originating in the sensory perception of an approaching shower, would be potentially vast and far too cumbersome for consciousness to make use of in the time available. Thus, although the whole galaxy would initially be 'lifted' from the memory archive, all its constituent items would encounter, on their progression towards consciousness down the elaborate electro-chemical conduits, increasing resistance until only a few crucially relevant data would actually get through to become a part of consciousness.

In this way, the flow of sensory impressions reaching the archival memory from the external environment would continually stimulate the archival memory to discharge almost instantaneously into consciousness all stored items of memory that were in any way relevant to the environmental situation that the sensory monitoring apparatus was then detecting. No other mechanism could explain either the immediacy or the magnitude of the flow of data issuing from each individual's archival memory.

Having reached this stage in understanding the operation of the brain, continued monitoring of my own thoughts revealed to me the further fact that everything that was in my mind at any given moment seemed to be derived either from the stream of sense impressions entering it from the environment or from the responsive stream of data issuing from archival memory. It is true that items from archival memory might then themselves reflexively evoke more items from the archive in what could become on occasion a protracted 'idea-chain' (day-dreaming, reflection, close analysis, solution-seeking etc) but the initial stimulant for the 'idea-chain' must invariably be an item of incoming sense impression.

I therefore came to realize that my mind apparently

consisted entirely (or at least so largely that I could not establish any other element with certainty) of a meeting of these two streams of data. The closest monitoring of my own consciousness of which I was capable simply failed to reveal to me any pre-existent entity (such as the true 'me' or 'my own mind') for the data streams to impinge upon and thus modify. I could only detect evidence of the flows themselves and of the turbulence generated by their meeting which, it became increasingly apparent to me, is actually what we call consciousness. At about this point I began to use the term 'twin data stream' to explain my developing theory of the nature of the human mind.

It is axiomatic that every moment (however fleeting a 'moment' may be defined as being) of a person's experience varies to some extent from the preceding one. Thus each passing moment will produce a modified sensory input which will then trigger the despatch from archival memory towards consciousness of a whole new galaxy of guidance data. This clearly suggests that the 'architecture' of the human mind must be very different from the relatively static model of it which is conventionally hypothesized. So far from being a fairly rigid structure holding at any moment a stable assortment of 'thoughts', the mind must in fact be an intensely dynamic and perpetually self-modifying confluence of incoming sense impressions and of data that has been summoned from the archival memory by the impact of the sense impressions upon it.

The turbulence of the meeting of these two streams must also perpetually generate new mixtures and combinations of thought so that it would perhaps be true to say, in a way analogous to what can be maintained about fingerprints, genetic fingerprinting scans, snowflakes and so on, that no two thoughts could ever be identical. The mind, then, so far from being relatively stable, is never the same from one 'moment' to the next. It perpetually modulates in phase with, and in response to, the individual's perpetually modulating environmental situation. There can therefore be no fixed 'programme' for consciousness analogous to a computer programme.

Another crucial inference implied by the responsiveness and adaptability of the system must be that human memory is not a passive repository for the storage of inert and unchanging stocks of data. It must rather be an intensely dynamic and context-responsive input-monitoring, despatch and, most importantly, data manipulating system.

This final quality of data manipulation derives from the fact that the stored data itself, as we shall later see, continually mutates under the influence of the perpetually varying impact of new data from the incoming stream of sensory impressions. The entire mind, in other words, is crudely analogous to a complex electronic assembly geared to a very sensitive environmental monitor. But, exhibiting a different order of subtlety and responsiveness to any product of human electronic technology, the configuration of every element of the brain-mind system would itself be in a continual state of dynamic modulation.

The following then is the model of the human mind that I had achieved at this stage: it is composed of an incoming flow of sensory data derived from hearing, smell, taste, touch and, above all, from sight. Moment by moment this stream automatically summons up complementary guidance data from a dynamic archival memory. These two streams meet turbulently at the point in time known as 'the present moment'. At any given instant the human mind consists of a dynamic swirl of what might be called data particles. These derive from the mixing of the data stream stemming from sensory monitoring with the data stream stemming from archival memory.

I am, incidentally, aware that I have been using the term 'flow' in a metaphorical sense. It is certainly a fact that complex electrical flows participate in the processes I have been describing but at the level critical for consciousness generation the key events are the perpetual activation, deactivation and reactivation of immense, continually modifying patterns of data cells probably distributed throughout the entire human cerebro-nervous system. For all that, it follows from the Twin-Data-Stream Theory that the mind itself, conceived as being the dynamic swirl caused by two data flows meeting under considerable pressure, must be in a permanent state of change.

For each individual mind, each instant of experienced reality is to some extent different from the last. The changing sense impressions of each passing moment automatically summon up from the immense, and specific to that individual, archival data stores hundreds, thousands or conceivably even millions of relevant items from the archival memory. However, only a small proportion of these ultimately get through the cascade of screens and become mind. Nevertheless, even if only a relatively minute proportion penetrates the screens the implication remains that each individual human mind changes partially but fundamentally from moment to moment.

This cyclic character of mind must operate, in order to cope with the rapidity at which the conditions of experience change, at a frequency of at the very least several times, and perhaps hundreds of times, per second. It seems to me highly probable that the oscillation of electrical activity is the same as the one which manifests itself macroscopically as the alpha wave (the basic brainwave of the encephalogram) which has a frequency of about ten cycles per second. In support of this supposition is the fact that the alpha wave is chiefly present when the subject is awake. It shuts down in sleep. Another, but far more remote, possibility is that the frequency of 'mind generation' reflects, and is causally linked to, the rate of the firing of nerve cells which occurs at a frequency of about five hundred per second.

The fact to be stressed in the present context is that the archival memory discharges a cloud or spray of data items towards consciousness in response to each summons from the data stream which derives from sensory monitoring of a perpetually changing environmental reality. This input would, if uncontrolled, constitute a siege by data that would be far more than consciousness could possibly handle. For this reason much the greater part of the data stream from archival memory is screened out before reaching consciousness.

The implied screens are probably arrayed in the form of a cascade varying from coarse to very fine, and thus only a tiny fraction of the data stream from archival memory ever actually 'becomes mind'. The fundamental evolutionary reason for the development of the whole extraordinary system has been to ensure that consciousness receives, at every new and unique moment of existence, a distillation of all the previous relevant experience of each individual in order to help him or her to manage encountered reality. In other words, throughout life each human mind has permanently at its disposal the very best counsel available.

The 'rogue' thoughts, which brought the entire mechanism to my attention in the first place, can thus be classified as irrelevant, or only slightly relevant, data items that have somehow failed to be screened out. They have passed, 'illicitly' as it were, through the entire screen cascade and reached consciousness. But so far from revealing a basic inefficiency of the system, the very fact of the rarity of such 'rogues' demonstrates a functioning information processing and retrieval system of extraordinary speed, flexibility and power.

This vision of the nature of the waking mind struck me as both awe-inspiring and rather alarming — alarming because the relentless dynamism and functionalism of the model seemed to clash fundamentally with the romantic image that most of us hold about the mind. This traditional, rather avuncular, notion implies a mode of consciousness which is — ideally, at least — calm, measured and autonomous. The Twin-Data-Stream Theory substituted for it something resembling that of a delicate winged creature buffeted continuously by opposing gales. But alarming or not, there seemed to me little doubt that this turbulent model was essentially correct.

# The Ceaseless Work of the Brain

It will be useful to start this chapter with a possible example of diurnal brain function. Imagine a man taking a walk. As he passes a dog the animal barks menacingly. The man's multi-sensory response to the implied threat is split into three identical streams. As mentioned in the previous chapter, one of these streams is transmitted to consciousness and one to the archival memory. We shall reserve examination of the important destination of the third stream until a little later.

On receiving the sensory impressions of the menacing dog, the man's archival memory responsively selects out and discharges towards consciousness every single item of data that is held in the memory archive that is even remotely relevant to the situation. It is hard to guess at the potential size of this output but clearly it would include items concerning dangerous animals, threats made by animals and especially dogs, techniques for coping with such threats, historical and literary allusions, the experiences of friends in similar situations, items derived from reading or other cultural sources, philosophical and legal considerations concerning rights and duties, medical implications and so on. Thousands, hundreds of thousands or possibly millions of data units could be involved.

This immense torrent of data is selected from archival

memory by the automatic retrieval system and despatched towards consciousness. But before arriving there, it passes through a screen cascade which eliminates perhaps 99.999% of it. The remainder, itself a substantial 'collage' of the most immediately relevant data, might then merely flicker virtually undiscerned in the conscious mind of the man as he walks past the dog.

An instant (perhaps a tenth of a second) later, a new, or partially new, environmental situation evokes a partially new torrent of information from the archival memory. All through each waking day this invisible and largely unperceived flow of context-selected data sprays, in one-tenth of a second bursts, from archival memory towards consciousness and a small proportion of it — the residue passed by the screen cascade — becomes mind. What is the relationship between this ceaseless diurnal generation and regeneration of consciousness and the nocturnal process of sleep and dreaming?

According to the 'twin-data-stream' theory of consciousness the incoming data flow of sense impressions is divided into three identical 'copies' on reaching the brain. Two of these, as we have seen, impact respectively on consciousness and on the archival memory but it is the third stream which is of special relevance to nocturnal brain function. In fact, this data flow impacts on a constituent of the brain/mind complex which has not yet been mentioned. It is a short-term memory which holds all sense impressions reaching the brain during the waking interval between periods of sleep. This interval will normally be about sixteen hours.

We must now relate this third data stream — the 'copy' that is shunted into a short-term memory — to what actually happens in the brain during sleep and dreaming. The purpose of the third 'copy' of the sensory input is to progressively build up a comprehensive 'stockpile' of the day's sense impressions for the brain to file and cross-reference during sleep. It is, in fact, this cross-referencing and filing of the sensory (and other) data accumulated throughout each day into the great archival memory which constitutes the immense nocturnal task of the brain.

The truth is that sleep, so far from being a time of rest when mental activity has largely ceased, is actually the period when the brain is performing perhaps its most impressive task. It is engaged in downloading from the short-term (sixteen-hour) memory into the long-term archival memory all the stored impressions of the preceding day.

Thus, the image of the dog passed briefly on a walk will, like all sense impressions, be stored in the short-term memory until a sleep period occurs. Then the multi-sensory images deriving from the experience will be downloaded into hundreds or perhaps thousands of data cells throughout the archival memory in obedience to exhaustive principles of relevance. The dog will be lodged in data cells about animals, about danger, about fictional dogs and other beasts, about relationships between dogs and people or dogs and other animals, in cells derived from reading, from anecdotes etc. Each impression that is downloaded from short-term memory will ramify into its own small reference library that will then subsist as a sub-branch of the immense archival memory.

Thus, when the man next passes a dog, the impressions that he derived from his archival memory on the former occasion will be fortified by new impressions specifically deriving from this new encounter. The earlier impressions will, if the next encounter is more than a day later, have by then been filed and cross-referenced into the memory archive. There will naturally be a broad similarity between the data items discharged by the archival memory in response to the second encounter with a dog and the new information obtained from that encounter. This new information will, in its turn, be lodged in the short-term memory and then committed to the archival memory the night after the encounter. In this way the archival memory perpetually transforms and enriches its stocks of data.

It is pointless to attempt even to guess at just how comprehensive the ultimate web of references will be. Nevertheless, what we can infer from everyday experience, which reveals the vast store of references that each new moment evokes (despite the great majority of them being screened out before they can arrive in consciousness), is that the net of the nocturnal filing and cross-referencing process must be cast very widely indeed. Just as the image of the dog is stored under hundreds, or thousands, of different headings, so is every other sensory experience that the brain receives during waking hours and consigns to temporary storage in the short-term memory system.

However, this data processing operation, immense though it is, represents only one part of the brain's regular nocturnal activity. Consideration of the problems of dynamic storage of perpetually modulating data suggests that the brain must also restructure each night the entire archival memory in order to accommodate new data, and facilitate later retrieval, in the most efficient manner possible. It is, in crude metaphor, as if a great library were not only given the task each night of housing hundreds of thousands of new volumes but also of cross-indexing these additions amongst the thousands of millions of existing books and finally of reorganizing the full range of storage and cataloguing facilities of the entire institution in order to provide users with easier and more rapid access to information.

My own realization that sleep and dreaming might, at the deepest level, be constituents of a storage, indexing and cross referencing system began many years ago when I perceived that one of the chief characteristics of dreams was the juxtaposing of incompatible elements in a way similar to that found in an index. One may, in a dream, converse with a crocodile or fly to a distant planet. One's home, as in the dreamlike *Alice in Wonderland*, may expand or contract. Shaking with anger, I once punished a person in a dream by beating him fiercely while he turned

slowly into a ball of wool and then began bouncing about the room. Any and every combination of imagery that can be produced by juxtaposing normally discrete elements from experience occurs in dreams.

And yet dreams are not fundamentally, or at least totally, irrational. They invariably include experience-related data. Although dreams, because of their bizarre imagery and illogical transformations often resemble hallucinations yet, to quote Polonius' musing on the apparent ravings of Hamlet, 'there is method' in them. The dreamer is often aware that while the content of his or her dream may be weird and distorted and the action defy all laws of logic or continuity the dream itself displays a perceptible, if obscure, relationship with experience.

As explained in the first chapter of this book, my own starting point for considering the nature of dreams was the observation that it is possible to experience two 'I's' in a dream. One of these 'I's' is an implied 'I' since it is never visualized but is merely the observer of the other. I also stated in the first chapter that research ultimately demonstrated to me that neither of the two 'I's' in 'two-I' dreams is the real 'I'. We can now see why this must be the case.

The fact is that the true 'I' — the living person — plays no, or only a very minor, part in dreaming. All dream imagery, including 'representations' of the dreamer, is in fact random imagery generated by the downloading and cross-referencing process which probably constitutes the major part of the nocturnal activity of the brain. The waking person (the true 'I') plays no part in dreaming and no elements of that person's mental existence or individual autonomy are necessarily present in dreams. This flat statement needs a degree of qualification: it is valid in the present context but will need elaboration later when we glance at one or two further modes of dream generation which must be given due weight.

The fact is that an image of the self that occurs in a

dream has no more connection with the dreamer's true self than an image of a kangaroo or of another person does. For this reason a dream may contain two 'I's' or a whole platoon of 'I's'. It all depends on the nature of the cross-referencing process that is taking place when the dream is being generated. Dream imagery is actually fabricated by fortuitous and automatic processes and in itself has no special significance. It might, in a tongue-twister-like but accurate way, be said that the dreamer does not really dream the dream — it is the dream that dreams the dreamer.

The human brain, scanning with its several senses, records into short-term memory *all* the experience that has occurred during a given day. Everything seen, everything heard, everything smelled, everything touched, every motor impulse and muscular contraction, every meaning-impregnated sight or sound (i.e. deriving from print or speech) — indeed, all that impinges on a person from within or without the body — all this is initially lodged for the remainder of the waking period in the short-term memory.

This extraordinary degree of comprehensiveness can be inferred from the fact that experience frequently summons from archival memory a recollection of a glancing, casual impression that was of no special significance when it was originally received but proves relevant when recalled. Such an event might be the recollection of a facial expression, a physical alignment — indeed virtually anything — which later, and sometimes after a span of years or decades, proves crucially significant or explanatory in the later context. It would thus seem to be *a priori* impossible that any principle of selection could have been operating at the time of the original input since any such principle would, of necessity, have had to be teleological and this, of course, is impossible. It is hard to see any possible explanation for the availability of such mundane impressions at a later time other than the storage of the totality of the sensory and other impressions that impinge on consciousness.

This 'total storage' may seem, at first consideration, an impossible feat. Large though the archival memory must be by any reckoning it can hardly store everything! Or can it? We know that each cell in the body actually does store the entire DNA code for reproducing a complete human being and this means that the brain is capable of storing data at the molecular level. We know too that the brain contains many billions (estimates vary wildly) of neurons elaborately cross-linked by chemically-mediated synapses into a reticular structure. Given this scale of data storage capacity, it is quite possible that the total sensory input for a day could, at a molecular level, be stored in only a few brain cells.

Such levels of storage capacity imply that every sensory and cognitive experience of a lifetime may be stored in some form, which could be digital or analogical, in the brain. It may also be the case that the short term memory is stored analogically while the long-term memory is held digitally. Were this true then a further, and again immense, part of the nocturnal activity of the brain would be converting the former to the latter.

I have invented a very short story to help illustrate the nature and mechanism of dreaming: 'Max Planck and Plato took some plutonium to the Pocono conference. Boris Podolsky was reading poetry there and playing poker with a polaron who had polio. John Polkinghorne was in love with a positron and found some antimatter in the Presbyterian Sanatorium.'

This unorthodox and alliterative little tale is, of course, nonsense but readers may perceive that it is not random nonsense but betrays certain organizing principles. In the first place, an alphabetic principle can be detected since many of the terms begin with the letter 'P'. This suggests that the tale derives at least to some extent from the index to a book. Then again the names and the nouns seem to hint at the world of science. In actual fact, the story is derived from the index to James Gleick's excellent biography, entitled *Genius*, of the American physicist, Richard Feynman. But the section of index I have used has obviously been tampered with by me in an attempt to create a kind of story.

I did this because I was seeking to create from the index a continuous narrative in a way analogous to that in which sense impressions are processed by the brain into dreams. The key difference, of course, is that dreams are primarily visual in character while my tale is, inevitably, given its genesis, verbal. But with this proviso the structure and above all the aetiology of the two narratives have a lot in common. Both dreams and my tale take as their starting point a sequence of quite disparate elements which nonetheless have a certain underlying thematic and structural unity. This unity derives in the case of my invented tale from the index to a book and in the case of a dream from the fact that it is, in part, a kind of fortuitous index to the events of the previous sixteen hours of waking life.

All data retrieval systems depend crucially on the use of an index. The clue that finally led me to understand that the two phenomena — dreaming and data retrieval — were linked, and indeed stemmed from, as it were, opposite ends of a continuous process of data manipulation, was my growing perception, when monitoring my own dreams, that events from the previous day could often, albeit in a form that was usually distorted and always heavily adulterated with other elements, be discerned in them. The dream thus constitutes a kind of surreal index to the waking experience.

A meeting with an old friend, for example, might lead to his or her appearance — although perhaps as a stranger, a rival, an animal or even an inanimate object — in the night's dreaming. I once dreamt about my anxiety at having lost an important bill and only much later realized that the dream had really been about a close friend of mine called Bill who had moved to another country. Having perceived that the previous day's impressions are recapitulated, although often in deeply camouflaged form, in the night's dreams I soon made another and equally significant observation. This was that the temporal sequence in which events occurred during the day was reflected in the times at which they resurfaced in dreams during the night.

By this I mean that an impression received, say, just before going to bed would not, or was not likely to, figure in a dream until late in the night and might easily occur just before waking. It seemed then that the sequence of the day's impressions was reproduced in the night's recapitulation of them. Naturally, I was able to observe this effect most often with events that happened late at night since these were the ones that appeared in dreams shortly before waking.

This book is essentially the record of a thought experiment. However, I realized while writing it that many of its assertions were susceptible to being experimentally verified or refuted. I therefore decided to suggest possible laboratory and other tests which could be used to confirm or negate alleged discoveries. The observation about the night's dreaming recapitulating the sequence of the day's experience provides an opportunity for possible experimental confirmation or refutation. Dream laboratories are now abundant and there should be little difficulty in making minute records of a subject's diurnal experience and then attempting to correlate it with his or her nocturnal dreams.

Dreams are, of course, incomparably more complex than my little tale derived from a modified index. Dreams are formed from sequences of recapitulated sense impressions fused with previously stored data and as such are far more wide-ranging in subject, and elaborate in structure, than data found in any book index. Nevertheless they do bear a functional relationship to the world of experience.

As regards origin, the chief difference between dreams and my story is that I deliberately provided the distortions in my story while dreams do not 'deliberately' distort anything. They achieve their fortuitous distortions simply by being the random by-products of what is essentially the chief data-ordering system of the human brain.

In the light of the preceding analysis, then, what would be an accurate definition of the phenomenon of dreams? Dreams are the fleeting glimpses sometimes obtained by the partially conscious mind of the filing and cross-referencing process that occurs unceasingly throughout each sleep period. These glimpses are only possible during the brief periods when consciousness is returning after such a period. During full sleep dreams cannot occur since the images generated by the brain's activity must impinge on at least a partially conscious mind to become manifest. In actual fact, during full sleep consciousness has been almost totally obliterated.

There has been much speculation amongst dream researchers as to whether people dream all night and only remember some (if any) of their dreams or whether they only dream for part of the night. The truth, as we have just seen, is that dreams cannot occur when the dreamer does not perceive them since it is the perception that is in fact the dream. Again the nature of the dream — i.e. what may be perceived by returning consciousness as a person awakes — is crucially dependent on when the person awakes and how long the awakening process takes. If he or she wakes in the middle of the night then the dreams being generated will concern events from the middle of the preceding day. And the same condition is true for any time of awakening.

Dreams are creatures of the twilight period between sleeping and waking. They can only 'occur' when enough consciousness is being generated to perceive them and enough downloading of sense impressions is still going on to provide the imagery from which they are woven. Many readers will probably have found that, in the half-way state between waking and dreaming known as dozing, it is quite often possible to perceive dreams forming and at the same time remain marginally aware of sensory impressions still reaching consciousness

Between the nocturnal cross-referencing and filing process and the diurnal period when consciousness is being generated there is a brief period (perhaps lasting seconds and rarely more than minutes) when both processes to some extent coexist. As waking consciousness returns so the nocturnal filing and cross-referencing activity declines. But while the two activities are both at least partially operational, during the brief period of awakening, consciousness may glimpse the cross-referencing activity and such glimpses are what we know as dreams.

The illogical and surreal quality of dreams derives from the fact that they are collages of random aspects of a cross-referencing process that by definition has little or no narrative coherence or continuity. But it does have some. The images 'perceived' by returning consciousness are not moving images but the very fact of sequential perception endows them with the illusion of being moving images. We experience them in a way similar to the way we experience a film in a cinema. They seem to tell what is just recognizable as a story — a weird, often discontinuous, invariably distorted but still marginally coherent story.

The vitally important question that arises from this fact is: why do we perceive dreams as a story when they are in fact fundamentally composed, like my modified index, of disparate and semantically discrete images. The answer to this question is crucial to an understanding of the evolutionary development of the human mind which makes it powerfully oriented towards the generation of narrative. But the question takes us into areas that are so challenging and also so fundamental to the nature of consciousness that it cannot be dealt with summarily at this point and we will have to reserve it for later discussion in a chapter of its own.

It is, however, appropriate to point out here that these mechanically generated sequences of essentially meaningless imagery which we call dreams have, for thousands of years, been an influential and often destructive force in human affairs. The effect of crediting dreams with supernatural qualities has produced results beyond calculation in history. There is no doubt that innumerable lives have been changed, and many lost, as a result of the grievous fallacy of believing that the strange content of dreams conveys, or may convey, secret messages. The very destiny of nations has sometimes rested on elaborate interpretations of dreams although, as we have seen, they are really no more than distortions of stored and, in this context, random imagery.

On a more positive note, it is also probably true that dreams will, once their true nature has been understood and their content related to individual experience, prove to be of value in a variety of diagnostic and other therapeutic situations. It is indeed not inconceivable that dreams really will prove capable of manifesting genuine prophetic capability although this will have nothing to do with the paranormal or the supernatural.

The truth is that sleep is not primarily a rest period although very probably the body is to some exent being rested. But the remarkable, and hitherto unsuspected, fact about sleep is that it is the time at which perhaps the most intensive mental work that any individual ever performs is taking place, athough this task is completely hidden from the conscious mind. It is also a fact that, intensive though it is, the nocturnal task may be equalled by the brain's diurnal role of generating mind.

The twenty-four-hour cycle of brain and mind activity can now be specified: the brain receives during the day multiple (corresponding to sensory input and also to endogenous monitoring systems) streams of data, including linguistic and written data. These incoming data streams impinge functionally on consciousness and also on the archival memory.

A third 'copy' of the data stream is divided into manageable sections and lodged in a short-term memory system which is just about large enough to hold a day's input. We can infer this estimate of its capacity from the fact that dire effects begin to take place with prolonged sleep inhibition (much more than twenty-four hours). The malign effects of sleep deprivation can lead, as we shall see in a later chapter, to 'waking dreams' and even to death.

At night, when we sleep, all these stored sensory, and other, impressions are downloaded from the short-term memory store into the long-term memory archive. This is equipped with an extraordinarily subtle and comprehensive cross-referencing and data processing system. It is very likely that every item of data stored in the entire memory archive is reviewed and realigned each night in phase with the input of fresh data. In this case, each human being would awaken every morning with a totally reviewed, and partly reconstituted, archival memory.

This data-storage and manipulation system, which I have called the archival memory, is perhaps the most impressive, as well as being the most characteristic, part of each human being's mental composition since it is the structure that makes every living person unique. It also makes all human culture transmissible and hence, in the final analysis, possible. In addition, the system must be responsive to the most minutely specific requirements for guidance at every waking moment of each individual's life.

We must now consider a matter of considerable importance to the validity of the Twin-Data-Stream Theory. This is the nature of REM (Rapid Eye Movement) sleep. It has been known since 1953 that virtually everyone spends part of each night in a sleep condition characterized by rapid movements of the eyes. Since, of course, the eyes are closed it may be wondered how this can be detected. Monitoring is performed by means of a special form of electroencephalogram called an electro-oculogram which can pick up the minute changes in the electrical field caused by the rotation of the eyeballs in their sockets.

Since the initial discovery of REM sleep, quite a lot more has been learned about its electrical manifestations although no researcher has convincingly suggested a reason for it. It is, however, known that during REM sleep the brain, judged by electroencephalographic traces, is in a state similar to that of being awake. The REM sleep period also seems to be the one responsible for the most abundant and most vivid dreams. REM sleep occupies about a third of normal sleeping time but this third does not all take place at once. It occurs typically in four sections during the night. These periods add up to some ninety minutes of REM sleep in total. The durations of the sleep periods devoted to REM sleep vary somewhat.

Now the obvious explanation for the rapid eye movements which occur during REM sleep is, as Empson states, that these motions are the movements that the dreamer makes in accordance with the scenario of his or her dream. If a sleeper dreams that he is watching a gull flying from the left to the right of his field of vision then his REM movements should reveal his eyeballs moving accordingly. But Empson rather despondently admits that attempts to demonstrate this correlation in sleep laboratories have thus far proved a failure.

My own view is that the rapid eye movements in sleep have nothing to do with — or at least nothing directly to do with — dreams. I am convinced that what the rapid eye movements occurring during REM sleep periods represent is a mechanical recapitulation of some or all of the eye movements that the individual has made during the preceding day. If this is true then it explains the well-known fact that blind people do not experience REM sleep. There are naturally no, or only random and few, eye movements stored in blind people's short-term memories that would be capable of being recapitulated in sleep. It is therefore very likely that the rapid eye movements are, in fact, merely a non-functional involvement of motor circuits with the nocturnal filing and cross-referencing process of the previous day's visual experience. REM sleep, in other words, is the time when visual impressions are being filed and cross-indexed into the archival memory.

A support for this hypothesis can be found in the fact that sleeping dogs that are producing rapid eye movements (as many animals do in sleep) may also make running movements with their legs. Dogs are known to dream and it seems reasonable to suppose that the reason why dogs make running movements and rapid eye movements is in a twin recapitulation of their waking activity. Later in this book we shall see that a similar phenomenon, sleepwalking, is quite common amongst humans.

A convenient experimental verification, or refutation, of this theory of the nature of REM sleep can be readily suggested. It would take the form of keeping voluntary subjects awake in a dark experimental chamber for a period approximating that of a full waking day. The room would be completely dark except for a continually moving beam of light that would be automatically recorded on film or tape. The experimental subjects would be asked to follow the moving beam of light with their eyes. During the subsequent REM sleep period it should be possible to make correlations between the REM motions of the eyeballs and the recorded movements of the light beam.

Another validation of the theory would be provided by the demonstration of some kind of equivalent of the rapid eye movement phenomenon taking place in another sense organ. The most obvious choice would be the inner ear. It might be possible to demonstrate a similar kind of motor involvement in the eardrum or the acoustic bones on downloading auditory impressions as that shown by the rapid eye movements in the case of downloading visual impressions.

This chapter has concentrated on describing the actual mechanism of nocturnal mental functioning. In the next we will look in detail at the role of the archival memory in generating dreams by night and supplying a data feed by day and how these two activities, one fortuitous and one functional, complement each other.

# Night and Day

Belief in the supernatural powers of dreams has been held by credulous people throughout human history. This credulity has itself generated another class of people claiming the ability to decipher the hidden messages of dreams. This second group has included not only shamans and other religious functionaries but also plain charlatans. Numerous books have been written relating the 'symbols' in dreams to future events. Astonishingly, fragments of such works survive from at least the second millenium BC.

The modes of interpretation used by most dream prophets are nothing if not simplistic. For example, if a sleeper dreams that he or she is travelling then the dream interpreter is very likely to tell the person that they will soon go on a journey. Prophesying by means of dreams is technically called oneiromancy and even today attention is still paid by scientists to its actual possibility.

Needless to say, dreams do not really have any supernatural or extra-sensory powers whatsoever. It is nonetheless possible that, once their real nature is understood, they may really be able to cast some light on restricted areas of the human future. Close analysis of a person's dreams might, for example, expose aspects of his or her mind that could not be made manifest by any other means. If the subject were an influential public figure, then this mildly 'prophetic' effect could conceivably extend to events on the world stage.

The truth is, however, that obtaining useful insights into the contents of a mind by examining a person's dreams is — as the psychoanalysts discovered long ago — fraught with difficulties. This is largely because, as we have seen, dreams represent only a very small and essentially arbitrary sample of mental content and functioning. But second, and more important, is the fact that insight into the origins of dreams can usually only be made by the dreamer him- or herself and even then only with great effort.

Each dream is so intimately entwined with such vast areas of the dreamer's life experience that no-one but the dreamer can perceive many of the elements that compose it. Sometimes a close relative — or, better still, a wife or husband — who has shared many years of life with the dreamer, can make some progress with the task.

There is little doubt, however, that dream content will ultimately be found to vary in consistent and probably classifiable ways according to such factors as education, lifestyle, profession, interests, nationality, normal or pathological brain states and so on. A true science of dream analysis leading to valid insights into the mental or even physical condition of patients may thus become possible. It has, after all, been known for a long time that discomfort registered in dreams (irritation or pain in various parts of the body) may give an early warning of the development of genuine medical or psychological problems.

Although dreams have no supernatural powers, the fascination which they have inspired in people at all times and in all cultures has given them an influence far in excess of their actual content. How many superstitious generals have won or lost battles because their strategy was influenced by a dream? At a more homely level, how many families or individuals have embarked on a course of action that ultimately proved disastrous (or beneficial) as the result of a dream?

There can be no doubt that world history, and probably most individual history, would be different (and different sometimes to an extraordinary extent) if the intriguing, but inherently arbitrary, wisps of surreal imagery known as dreams were not intermittently produced by the ability of the mind-generating brain to gain occasional glimpses of its automatic sleeping task.

In an earlier chapter, when explaining the mechanism of the 'two-I' dreams which first set me on the path to writing this book, I stated without qualification that the 'I' (or multiple 'I's') that figure in dreams cannot be identified with the true, or waking, 'I'. I must now — partly because of a dream record that I kept while writing this book — modify this statement somewhat. Out of a vast number of theories concerning the origin and purpose of dreams, there are, in addition to the Twin-Data-Stream Theory, two which we must look at. One of these is the largely proven theory of dream 'incorporation' and the other, which overlaps with the first to some extent, is Sigmund Freud's theory of the function of dreams.

Dream incorporation is the process by means of which some stimulus, and usually a disagreeable one, experienced by a sleeping person may become incorporated into a dream that is actually taking place at the time. There is no doubt that such incorporation does occur although, as Empson states, usually only in an indirect and often obscure way. For example, subjects in dream laboratories who are kept hungry or thirsty rarely report, as they might be expected to do, dreams of eating or drinking. Again if water is splashed on a sleeping subject this does not necessarily cause that subject to dream of getting wet, although it may do.

I have myself, however, experienced the incorporation of fortuitous stimuli into dreams often enough to be convinced that the process really takes place. Pain, hunger and other bodily irritants quite often find their way, most often in distorted form, into my own dreams.

Freud believed that the chief function of dreams was to protect sleep. An alarm from within or from outside the body (a headache, a doorbell ringing) that threatened to wake the subject would cause him or her to fabricate a dream in which the problem was dealt with. Thus, in the dream, a pill might be taken to relieve the pain and the doorbell could be answered. The subject's awareness might thus be deceived into supposing the irritant had been banished and sleep would continue.

Both dream incorporation and sleep protection are superficially incompatible with the Twin-Data-Stream Theory since the latter maintains that dream production is exclusively dependent on the automatic functioning of a filing and cross-referencing system in the sleeping brain. The resolution of this apparent contradiction may well derive from the fact that the true 'I' is never completely disabled during those periods of sleep when dreams are taking place.

On the contrary, it is a crucial aspect of the Twin-Data-Stream Theory that dreams can *only* be generated during the twilight period when the mind is coming back into existence and the downloading of sensory impressions into archival memory is coming to an end. Dreams are not formed during those long periods of the night when the mind has been 'switched off' and the brain is operating purely as an electro-chemical data processing device.

The twilight period between waking and sleeping involves a gradation of intensity as regards returning consciousness. During this period it is possible that the dream 'I' and the true 'I' might achieve a kind of fleeting congruence. The relationship between the dream 'I' and the real one is clearly complex and graduated. It is therefore possible that a stimulus received by the sleeping body might partly waken the person and introduce a perceptible degree of volition into dream formation. This would probably result in an elaborate interaction developing between the partially awakened consciousness and the mainly automatic process of dream formation. In such circumstances the archival memory might be stimulated to supply imagery appropriate to the irritant.

It is even possible that such a sleep protection system has evolved to the point where it is by now a true function of archival memory. The mechanism historically involved in bringing about such a development is usually a form of 'tinkering', that is cobbling together new systems and devices from elements of old ones. Evolution can and does work to some extent by means of the tinkering process. It is, of course, every bit as important for sleep to be guarded in terms of the Twin-Data-Stream Theory, which postulates sleep as a period devoted to intensive, and essential, data processing, as it is in the context of classical theories concerning the body's need for rest.

It is, for example, undoubtedly the case that dreams have always interacted at a profound level with aspects of human culture. Their production may be fortuitous but they nevertheless constitute wondrous mental and spiritual adventures. These can open a perspective on events that is complementary to, and quite as revealing as, waking experience.

In this context, it is worth recalling that Arthur Koestler called his study of great scientists and their achievements *The Sleepwalkers*. Koestler maintained that scientists do not in fact achieve their most profound insights by means of targeted programmes of research but rather as a result of almost visionary revelations. The suggestion behind Koestler's thinking is that the kind of associations and connections that are formed in dreams are far more likely to give rise to perceptions that help us to understand and explain the cosmic environment than any amount of meticulous experimentation in laboratories.

The meaning of dreams — and I mean their origins in experience and not their alleged supernatural significance

— can often be teasingly hard to detect. As a testimony to this fact, I will begin my own dream record by referring to a brief passage in a recent dream which demonstrated how deeply camouflaged experience may be after it has been subjected to the subtle cross-referencing procedures of the archival memory.

This dream ranged widely but I was able to isolate and refer to source many of the elements within it. I found, however, that there was one perplexing stretch of it that I could not relate to any 'trigger' experience. This was a brief period in the dream when I found myself on horseback and what is more I was a proficient and daredevil horseman. In reality, I can hardly ride and was, in childhood, once bucked off a horse. But in the dream I galloped proudly up and down, feeling that everyone was admiring my horsemanship.

I combed my activities of the previous day but for a long time found nothing that might have inspired the equestrian element. But finally I located the inspiration for it. I had lunched with an old friend whom I had not seen for years. Over our meal, we both became slightly nostalgic about the past. She had talked rhapsodically about her very happy childhood in Yorkshire. And as I recalled this fact, I suddenly realized where the mysterious passage in my dream had come from.

I never think of Yorkshire without thinking of the curious and poetic name used for its major administrative districts, the 'ridings'. The moment I remembered this I realized that my friend talking about Yorkshire had put into my mind (without my realizing it) the word 'riding'. This word, although it had never actually been spoken over our lunch, must nevertheless have been evoked in my consciousness. It would then have been lodged in my short-term memory and that had been enough to set me galloping and bucking about on a horse in the early hours of the following morning. The 'bucking', incidentally, is doubtless testimony to the fact that my archival memory, seeking data cells in which to lodge the subjective experience of the word 'riding', had located some that held memories of my childhood experience of being bucked off a horse.

Since I started work on this book I have attempted to keep a dream record. My efforts have not been very successful, partly because of the well-known difficulty of recalling dreams even moments after awakening and partly because, lacking experience, I ran into technical difficulties. My way of gathering and storing my dreams was to use a micro-recorder which I kept close to my bed. However, I often forgot to place it in the right position and, on awakening, had to fumble for it while the dream memory rapidly dissolved. Then again the micro-recorder sometimes switched itself off.

Because of the inadequate and fragmentary record of most of my attempts to harvest dreams, I have decided that I will restrict the present transcription to just two dreams, the first and the last that I recorded. More or less coincidentally these two represent my least and also my most successful dreams in terms of illustrating the process of dream formation as specified by the Twin-Data-Stream Theory. The first dream evoked nothing whatsoever that I could recall specifically while the second one teemed with references. Here then are the accounts of these two dreams, with my fragmentary comments, recorded just after waking.

'What was I dreaming? I don't know what I was dreaming. There was a sea front — the only thing I do remember is — we were clearing ice — and I found the ice was five or six feet deep — there was a gully — and then someone — a girl — a woman — arrived to help — and I don't know who it was — and from the edge of the sea a sort of parade, as at Brighton, or a sort of pier or esplanade — and the depth of the ice took one by surprise. Almost as soon as I knew the job was ice-clearing the ice turned out to be very deep. And, of course, one knew this because the ice had this slash or trench in it and I jumped into it quite easily. But where was I? I don't know.'

That is a verbatim account and I have no real notion as to the derivation of any of it. I could relate its imagery to 'Freudian (erotic) symbolism', to aspects of my experience whether recent (in relation to the time of the dream) or in the remote past but I have no idea if such associations would really reflect the actual aetiology of the dream. I have, naturally, often been to the seaside, walked on esplanades and cleared ice, especially when I lived in America. But these images do not suggest to me any remembered sense impressions of the previous day.

There is one depressingly self-satisfied suggestion I could make as regards the origin of the dream and this is that it derived to some extent from my own work on dreams. Were this the case, the dream would include an allegorical representation of my attempts to get rid of the frozen state of dream research, i.e. 'clear the ice'.

It is, of course, a fact that dream formation, like literary creation, generates allegory, metaphor and symbol. But the inference I have made above does not derive from remembered experience and therefore cannot be credited with any degree of objective likelihood. Dreams are frustratingly hard to interpet not only because of the enormous spread of potential references that they contain but also because their structure fuses a naturalistic and a surreal element.

The naturalistic component is narrative progression. The dream moves from episode to episode in a way analogous to that in which a novel or play proceeds. The second element consists of the narrative's bizarre and naturalistically impossible content in which both events and material objects metamorphose at a speed and in ways that have no counterpart in reality although the metamorphoses are not dissimilar to those seen in some videos.

These two structural elements reflect the process of dream formation as specified by the Twin-Data-Stream

Theory and thus tend to confirm the validity of the theory. The logical progression from scene to scene derives from the archival memory filing and cross-referencing impressions received in sequence from the short-term memory. The surreal content represents the huge array of data cells into which each impression is cross-referenced and filed.

The next, and last, dream that I will attempt to analyse here is the most useful one in my entire dream record. In fact, after having experienced the following dream I made no further attempt to capture dreams since this one supplied almost everything I had hoped to obtain from keeping the record. Oddly enough, the dream's chief virtue stems from the fact that it is superficially a drab and unexciting dream. But this makes it an excellent one with which to test the value of the Twin-Data-Stream Theory for explaining the origin of dreams.

'In this dream, which seemed to go on for a long time, I was a soldier. I, and some others, were garrisoned on a kind of concrete platform. I owned both a motorbike and a car and I went away on a trip that had been organized, or at least permitted, by the army. When I returned I found difficulty in getting both the car and the motorbike back to base. This difficulty was of the kind familiar from puzzles or paradoxes, i.e. how to move both motorbike and car across country with the minimum of travelling back and forth.

'On my return, I found that the concrete platform was now almost submerged and water lapped over it. I wondered how dangerous this might make it as a place to live. Then I realized that both my vehicles, car and motorbike, were falling apart and I was distressed because they were rented vehicles. Then I lost track of both of them and realized that they'd been impounded by the army.

'An officer wearing a long trench coat arrived and began to question me. He said that the motorbike had been found but with some money and some pills in it. I asked myself: "What have my vehicles been used for?" The officer was suspicious and asked me in a faintly sinister way: "Are you going to be here for a long time?" I then realized with sinking heart that I was going to be in the army for a very long time and I said: "Yes, I suppose so, sir." And he said: "Right."

'I thought to myself: "Perhaps he will now explain to me what exactly is going on." But then I saw that a round hatch, a bit like the conning tower of a submarine, had opened at his feet and liquid was bubbling in the opening. The officer slowly descended into this hatch as if on a stage lift. I wondered where he was going. Could there be living quarters or workrooms down there? I feared he would drown because there was a lot of water in the opening — but oily, soapy water. The officer sank down until he disappeared.

'Later, a sergeant arrived with my motorbike and said: "There's something funny about this." I saw that he was wheeling the machine with difficulty and could only move it a small distance. I wondered if this might be because there was a bomb attached to it. There were now five of us in the bunker which had acquired a very low wall and I said: "well, if that bomb's going to go off I think I'd better lay down on the ground." I also said: "It's not going to be very healthy for us if it does blow up — we'll get glass all over us." But the bomb did not blow up and the officer reappeared. And at about this point I woke up.'

When recording the dream on awakening, I interpellated comments such as: 'It was a complex, structured, highly-evocative, highly itemized, cross-referenced dream and there were a lot more references to yesterday evening's events than I can . . .' And at this point the recorder switched itself off. My excitement grew and later, having coaxed the recorder to work again, I interspersed the dream record with remarks like 'Oh, this is a marvellous dream' and 'Oh, it's a goldmine, this dream' and I concluded with the wondering remark: 'There may be thousands of references condensed into that little sequence of dream.'

I will now excavate some of these references. The point of this is to show that the most bizarre and unlikely elements of dreams derive from past experience and that their derivation can quite often be demonstrated.

I have served in the army and made various journeys as a soldier. These bare facts are, I feel, inadequately specific for me to be able to claim them as purposeful dream references. However, the puzzle or paradox of how to transport both my car and my motorbike across country almost certainly derived from my recent fascinated reading of Paul Davies' book *The Mind of God.* This work stimulated my interest in quantum theory, Zenoesque paradoxes and especially the 'liar paradox'.

In addition to the 'paradox' aspect to my concern about moving both the car and motorbike there was a more practical one. My younger son and I had recently discussed the possibility of using a car to carry tents and heavy luggage while walking the length of Hadrian's wall. We could not, naturally enough, figure out any way to make use of the car without having to walk the wall twice.

There were undoubtedly a number of references to the parking situation in the block of flats where I live. One of these was the concrete apron on which the military encampment was originaly located. Then there has been a lot of controversy in the building recently about thoughtless tenants who park vehicles, cars and motorbikes, without authorization. There are often motorbikes in various stages of disrepair lying about. I rent and partly sublet a car-bay and this was probably the origin of the 'rented vehicles' reference in the dream. The vehicles being 'impounded' may well be a reference to the continual threat of having one's car removed or immobilized.

My family was about to leave for a long weekend of camping in Kent. When trying to decide where to swim we looked at photographs in guide books. Some of these showed a resort which had stone steps leading down into the sea. Doubtless this stepped concrete esplanade was, along with the concrete car park, in part responsible for the image of the concrete army camp in the sea. The forthcoming trip to Kent would also have generated the reference to the motorbike being returned with pills and money in it since pills (paracetamol etc.) and money were two of the items on our list of things needed for the trip.

The army officer, I realized after some consideration, was probably a version of myself, a second 'I'. I did not at first recognize that the greatcoat the officer wore in the dream was one that I had once owned. The officer did not, in fact, look especially like me but in addition to the coat there were other elements in his appearance which convinced me that he was indeed a version of me. The reference to the anticipation of being a 'long time in the army' is a recurrent one in my dreams. It often seems to me that I have frequently been banished to institutions, boarding schools, the army and the like where I was homesick.

The round hatch which opened at the officer's feet proved to be a fascinating amalgam. It almost certainly included at least the following three elements: Muraroa Atoll, an oil rig that was recently prevented by Greenpeace activists from being sunk at sea, and a broken compass. Muraroa Atoll is, many readers will recall, where the French, at the time I am writing, have expressed their intention of conducting underground nuclear tests. It is a coral island and the test chamber is, I understand, reached via a hatch in the coral. I read a description of this subterranean blast containment chamber and feel certain that it forms part of the 'conning tower' image.

The oil rig also had a structure reminiscent of a hatch poking up out of the sea and I have no doubt that a reference to it was included in the dream. But as a glimpse of how detailed the cross-referencing net of the memory archive really is, the third association is the most striking.

It is also the one that I thought of first when recollecting the dream. It is the image of an orienteering-type compass which I had owned for some years. Recently, on a country walk, I had accidentally broken it.

I discovered this breakage in two stages while my son and I were walking. First I found, after having crossed a stile, that some oily fluid had fallen onto my hands. It was most unpleasant. I wondered if it might be animal urine or some kind of — possibly corrosive — chemical. I tried hard to clean my hand with grass but could not remove all the oily fluid. Later in the walk I tried to consult my compass and found, to my surprise and dismay, that it was broken. The transparent cover of the needle capsule had fallen off and the fluid in the capsule had poured away.

It took me a moment or two to grasp that the compass fluid must actually have been the oily substance I had found on my hand. The compass must have swung against the stile as I crossed it and cracked open. When I ruefully mentioned this to my son we both laughed at the ludicrous mishap. But my own laughter was tinged with regret. I had become attached to this little compass which had guided us on many walks over the years.

The officer, who was a version of myself and who apparently sank down to drown inside the conning tower hatch, may have been symbolically drowning in grief both in, and because of, the lost compass. Dreams often, on analysis, reveal the strength of hidden emotional states.

The bomb discovered on the motorbike was a reference to an occasion, years ago, when my wife and I were touring in Ireland during the troubles and, while in Dublin, noticed a curious metallic box apparently strapped to the underside of our car. Suspecting a bomb, we notified the police. They found that the 'box' was simply part of the car's engine. Naturally we felt rather foolish about this. The dream bomb doubtless also included a reference to the Muraroa Atoll atom bomb.

That is all I can discern with reasonable confidence

about the iconography of this dream. However, I was aware when I first awakened and recalled having dreamed it that a vast number of other elements from past experience were bound up in it. But by now it would simply be guesswork to attempt to excavate more of them.

Still I think I have listed enough to give some hint of the scope of the data processing activity of which dream formation is a by-product. The dream as outlined above shows that my brain, at the time, was cross-referencing and filing data that had originated over a time-span of at least three decades and that ranged in subject matter from a suspected encounter with terrorism in Ireland to a broken orienteering compass. This gives at least an indication of the task performed nightly, and ceaselessly, in sleep by the human brain.

It is, of course, the case that the brain's nocturnal task has close affinities with its daytime work of generating consciousness and mind. The aspect of its diurnal activity which closely resembles nocturnal dream formation is the despatch of data from archival memory in response to the impact upon it of sensory impressions from the outside world. In both the diurnal and the nocturnal procedures, consciousness, whether it is partial or full consciousness, receives information that has been stored in the memory archive.

In the case of dream formation the relevant data is in the process of being stored and is only glimpsed more or less 'accidentally'. During the day, however, the data flow is despatched purposefully from archival memory to help consciousness at each experiential moment. But the two cases are similar in that the data received by consciousness in either case may be obscure in its origin and mysterious in its purpose.

I will give an example of the diurnal process at work since it reveals the cryptic methodology of the waking brain in the same way that the dream about 'riding' showed this quality in the sleeping brain. It also demonstrates that dreaming, dependent on purely mechanical processing, and thoughts issuing from the archival memory — thoughts which are, of course, context related — both use imaginative, metaphorical and even literary (story telling) abilities that are found at all the stages of data-manipulation. This account might be entitled 'The Anatomy of a Fleeting Thought'. Indeed, so fleeting was the thought that I would probably never have registered it at all had I not, as a result of being occupied with this book, been sensitised to such occurrences.

Having parked my car, I walked to, and then entered, the block of flats where I live. As I moved into the small foyer, I noted that our regular cleaner was polishing the BRASS panel of the entryphone system. His back was towards me and we did not exchange greetings. I walked the few steps to the lift. The lift car was there and I opened the gates and entered. It was at this point that I became aware of the faint image of a restaurant in my mind. Looking back, it seems to me quite likely that the image was held only for a single manifestation of consciousness, a tenth of a second, and that this period was just sufficient to enable me to catch the image 'on the wing'.

Even as the lift doors closed behind me the image vanished and I forgot it — but not quite. As a result of my apprenticeship in spotting 'thoughts that come from nowhere' I made a conscious effort to retrieve what I had seen faintly in my mind. Then, having succeeded in doing so, I stood motionless for a moment while some such thought as: 'But why on earth did I think of that?' formed in my thoughts. I then pressed the button to ascend and, as the car began to rise, stood concentrating hard, trying to determine why that particular faint, unassertive image should have been offered to me by my archival memory at just that moment. I had recognized the restaurant the instant I perceived it.

It was a famous, left-bank, Paris restaurant which I had

passed almost daily during the period when, as a young man, I had spent a year in the capital of France. I had also dined in this restaurant a few times, most recently with my wife perhaps ten years previously. As I rode up in the lift, I was still able to summon a ghostly after-image of the establishment including the broad Boulevard St. Germain, the glowing but plain interior and the big illuminated sign. It was undoubtedly the BRASSerie Lipp.

As mentioned, I asked myself: 'Yes, but why?' and immediately perceived at least part of the answer. The cleaner in the foyer had been cleaning the BRASS panel of the entryphone system. So what more natural than that the complementary image of the 'BRASSerie Lipp' should have been offered to me by my archival memory? Then I became aware of a query forming in my mind. After all, the fact was that neither the word 'BRASS' (which may have come fleetingly into my thoughts when I saw the cleaner at work) or the metal 'BRASS' that I had actually glimpsed on the entryphone panel seemed to explain why the BRASSerie Lipp should have been offered to me as, presumably, a natural response to the perceptions.

I knew that I could, given the time, have listed a hundred, or perhaps a thousand, mental associations with BRASS — both word and metal — which the archive might, just as usefully, have selected. It had taken me a minute or so after passing the cleaner to reach the lift and yet all that had come to me in that time was a faint impression of the BRASSerie Lipp to testify to the Herculean task of data sifting that must have filled this period.

Thinking along these lines, I came to the conclusion that my archival memory, and later the screens which had 'passed' the image of the BRASSerie Lipp, and despite the enormous effort that must have been put into the attempt to supply me with a useful reference, had simply misinterpreted my situation in space/time and the needs engendered by that situation. Surely, the recollection of, say, the BRASS buttons that I had wearily had to polish in the army would have been as useful (or useless) to me at that moment as the image of the Paris restaurant?

But then a little cluster, or constellation, of further references that might have informed the archive's 'choice' began to form in my mind. As a result, and after a little more reflection, I began to perceive that although the archive and the screens might have failed to supply a useful reference on this occasion their selection had by no means been random or arbitrary as it would certainly have been had I been offered the buttons. On the contrary, the image of the restaurant had actually been the consequence of an astonishingly close survey of fairly obscure aspects of my recent experience. The following are, I am pretty sure, a few of the reasons why the archive ultimately offered me that quick peek at the BRASSerie Lipp.

The cleaner had been polishing BRASS. My wife often polishes BRASS candlesticks when we hold a dinner party. On such occasions, we often consult cookery books. But we also occasionally use them when dining alone and feel like having a specially good meal. We had consulted a new cookery book a night or two before preparatory to a gala dinner on the present evening. I had suggested that we might make a particular dish for which it gave a recipe. This dish was a rather grand version of pork sausages and sauerkraut. My wife had vetoed the suggestion because, as I had forgotten, she is not very fond of sauerkraut.

She may well have also vetoed a dish including sauerkraut when we had dined in the BRASSerie Lipp a decade or so before. I cannot remember but were this the case then a memory of her refusal would have been held in my archival memory. Whatever the truth about that, it is certainly the case that the BRASSerie Lipp makes a speciality of splendidly prepared German dishes, which often feature pork sausages and sauerkraut.

In brief then, the anatomy of this 'fleeting thought' would be something like: I arrived home, anticipating a specially good dinner, to find someone a bit like a restaurant doorman, such as the BRASSerie Lipp has or had in those days, in the foyer of my block of flats polishing a BRASS panel. The idea of 'BRASS', either from the metal or the word, had thus been channelled into my archival memory.

The archive then 'asked itself' which of the thousands or millions of data items it holds that in any way concern either the metal or the word BRASS would be most useful to me in the developing situation. It would have spent a long minute scanning all its files about my recent and past activities, including the polishing of the entryphone panel, the earlier polishing of the BRASS candlesticks and my very recent discussion with my wife about a sauerkraut dish of the kind featured by the BRASSerie Lipp. Finally and impressively, if also uselessly, it would have 'concluded' that the item about BRASS held in its data banks which was most broadly relevant to my inferred environmental situation was an image of the BRASSerie Lipp.

The implied message from the archival memory might have been: 'Now you mustn't forget the time when you and S. dined in the BRASSerie Lipp. She told you then that she didn't like sauerkraut. So whatever you have for dinner tonight don't suggest sauerkraut.'

All day and every day, minute by minute, tenth of a second by tenth of a second, the archival memory and its attendant screens perform feats of this kind. They are capable of misinterpreting the evidence but, responding to the best picture of our environmental situation that they can form on the basis of the information available, they continually select out from a vast library of references held in the archival memory the most appropriate guidance that they can find.

# The Mind of Man

There is a puzzle about dreams that perplexes academics and ordinary people alike. It can be called simply 'non-memorability'. Why is it that only a tiny proportion of dreams can be recalled and even these usually only with great effort? Dreams seem to dissipate like blown smoke at the very instant that our mind makes contact with them.

As we have already seen, dreams consist of glimpses of the filing and a cross-referencing process that takes place in the archival memory during sleep. They are generated during the twilight period when returning consciousness briefly coexists with the downloading of sense impressions. The visual quality of dreams is analogous to that achieved by riffling a pack of Tarot cards, or indeed any cards having varied pictures, in front of the eyes. The changing image that results has a certain degree of thematic unity and illusory movement but actually consists, like cinematic film, of a sequence of static pictures.

Although dreams can only form during the 'twilight period' between waking and sleeping this fact does not explain why they are so evanescent. The true reason why dreams elude the grasp of consciousness is that they are not, at least when first experienced, an actual constituent of consciousness. They are rather shadows glimpsed out of a corner of the mind's eye.

Their apparent thematic unity (which is itself tenuous) is a by-product of the cross-referencing process. When a given sense impression is cross-referenced into a great many data cells the resultant images are bound to manifest a slight degree of thematic continuity. The fact that the similarities between dream images are so minor stems from the fact that the paths of meaning connecting them may vary wildly. This is because even when the same image is being processed, its multiple destinations can include a large number of different data cells.

It is, for example, relatively easy to see how the sense impression of a pencil might be lodged in a data cell containing an image of another long, slim object — say, a minaret. It is less easy to perceive immediately the rationale of the PENcil impression being filed, in obedience to linguistic prefix-similarity rather than shape, in the same file as an image of a PENtagram. A dream formation derived from such a thematic linkage might therefore include a scene in which someone, perhaps the dreamer and perhaps not, is writing with a pencil that turns into a minaret while the subject is within a pentagram.

The diurnal use of the same imagery for mind formation will be far less surreal than the dreams since the screens will then filter out all but the few images which are most immediately relevant to the subject's life situation. Thus while the thought, or perception, of a pencil will doubtless evoke things like pentagrams and minarets these secondary images will in most life situations be filtered out by the screens before they can reach consciousness.

A dream is inherently dynamic. Its image must be changing perpetually and almost inevitably irrationally. This is because the dream does not consist of a single glimpse of a single sensory image being downloaded but is actually the scanning of a continuous process of filing and cross-referencing. It is possible sometimes on awakening to recall the strange and apparently irrational transpositions of location and other elements of dreams but these seem strange only to the waking mind. While the dreams are actually being generated they engender no sense of being odd or bizarre.

The reason for this is that while some degree of consciousness must, at the time when dreams are being generated, be present in order for them to be perceived, there is not yet enough of it (virtually no stream of sense impression data) to summon up any complementary interpretive data from archival memory. If the data stream from archival memory were fully functional it would mean that dream generation would have ceased. Thus if someone dreams that they are, say, married to a talking tree no comparative or corrective data will be evoked from past experience to refute the possibility.

Thus, for the duration of the dream the person will, in fact, from every available perspective, be married to a talking tree. The union will have been solemnized in a universe consisting exclusively of juxtaposed, but not necessarily related, fragments of past experience. In such a universe — the universe of dreams — literally anything is possible and all things are equally natural.

A static image is not a dream. For example a statement like: 'I dreamed of a doll that had Stalin's face,' is unlikely to derive from a real dream although it could perhaps be an extrapolation from one. But by being isolated from its parent dream the static image actually distorts the true nature of dreaming. 'I dreamed I held a doll that had Stalin's face which began to weep, changed into the face of the Virgin Mary and began to shine like the sun causing trees to sprout rapidly in a garden beneath it' is the kind of passage that a dream might generate.

Dreams usually seem to the returning consciousness which generates them to be evanescent and elusive — and so they are. They are actually no more than linked images whose production is as mechanical as the mechanized handling of materials in a factory. Even as a mind reaches out to grasp them they are whisked away. You cannot, except very occasionally, 'hold' a dream in your mind because it is never actually 'in your mind' at all. Once you have forgotten some fleeting glimpse of a dream it is lost forever.

Just occasionally, however, when you first perceive a dream, you may inspect it for long enough and/or wake up swiftly enough for it to shift from its natural state and become a downloaded item in archival memory. Then at some later time when you 'remember' this particular dream what you are really remembering is the image of it which has by then been filed and cross-referenced into the archive. Henceforth it will have durable existence and respond like other stored data to being recalled by the incoming sensory data stream.

So if, at a later time, you recall that you once dreamed that you were married to a talking tree you will probably smile with wonder at how you could ever even have imagined such a ludicrous thing. This is because you will now have at your disposal the vast stock of data concerning marriage that is contained in your archival memory and this will provide you with a true perspective on the notion of being married to a tree, including a realization of both the absurdity and the impossibility of it.

A dream, then, is the uncertain perception by returning consciousness — and the perception is to some extent generated by the returning consciousness itself — of a slow coil of random imagery. This coil or sequence of images has been rescued from the instant annihilation which the vast majority of potential dreams undergo by the fortuitous circumstance of its having been processed just as consciousness was starting to return. At this point, returning consciousness began to pick up the images, and this, along with the attendant slowing down of the data processing mechanism, permitted a dream to form.

But why is it, readers may wonder, that the 'sleeping mind' does not go on making contact with this filing and cross-referencing system throughout the night and thus go on generating dreams? The fact is that there is no such thing as the 'sleeping mind'. The very phrase is a contradiction in terms. The astonishing, and possibly to some people disturbing, truth is that in sleep the human mind is not just 'asleep' but has ceased to exist. What we call sleep is actually a period when the individual mind is no longer being generated. It has simply been 'switched off'.

Human beings are not, as we seem to ourselves to be, continuous entities. For about a third of our lives we have no existence. It is obvious that, considered exclusively from a physical point of view, we are still alive. Our bodily functions continue to operate. We breathe and our hearts beat. But our brains no longer generate our minds. It is normally supposed that a sleeping human being is much the same as a waking one although his or her mind is now in a state we designate as 'being unconscious'. The truth is very much stranger than this. A sleeping man or woman no longer has a mind.

Some readers may be distressed to learn that people are not continuous entities but can be, and every sixteen hours or so actually are, switched off like a light or a television set. It must be accepted, however, that in REM sleep, and also to a greater or lesser extent in other kinds of sleep, a person is not merely unconscious but has been switched off (although the switching off process has been performed by a sub-system of the brain itself). The person, moreover, has been switched off in exactly the same way that a light or a television set can be switched off.

In fact, the analogy with a light or a television set is a fairly exact one for these man-made artefacts both consist of a tangible item of hardware, the lamp or the set, and an intangible product, the light or the more highly organized image on the TV screen. When the appopriate circuits in the hardware of these items of manufactured goods are powered by an electric current then the intangible products, the light and the television picture, are produced.

The brain too consists of 'hardware', the mass of interlinked neurons which constitute its physical being, and an intangible product, the human mind. When the proper circuits in the brain are powered then the intangible product, the mind, is produced. But during sleep the mind-generating circuits of the brain are 'switched off' and the mind ceases to exist just as the light and the television picture cease to exist when they are no longer powered by an energy source.

Human sleep is, in fact, a state in which the circuits that generate the mind during waking hours are either totally unenergized or else are appropriated for other tasks. In either case there will be, throughout the sleeping period, no circuitry left in the brain that is being used for generating consciousness. Personality, character, individual history — all the attributes that define a person — will have been, for eight hours, abolished. You switch off the television and the screen goes dark. You go to sleep and the mind goes dark. Popular metaphors often contain more than a hint of scientific truth. It is certainly the case that the saying 'he went out like a light' is a technically precise description of what happens to all of us when we go to sleep.

Can it really be true that a 'person' is no more substantial than the glow produced by the white-hot filament in a light bulb or the more elaborate but wraith-like picture on a cathode-ray tube? Are human beings merely the intangible product of elaborate electro-chemical circuitry? The answer to this question has to be an uncompromising 'yes' although there may be, for those who find the notion hard to accept, a little consolation to be found in the probability that in non-REM sleep, and perhaps even (in an extremely attenuated form) in REM sleep too, a kind of phantom caretaker (a true 'ghost in the machine') continues to preside over the destiny of the body.

The purpose of this shadowy residual person would be chiefly that of ensuring the possibility of rapid revival of the extinguished consciousness in the event of some physical threat. It is remotely possible too that the phantom might also retain some rudimentary executive or decision-making powers. In REM sleep, however, the situation is likely to be, as the frivolous saying about acquaintances with lightweight intelligences puts it, that the lights are on and the door stands open but there is no-one at home. In REM sleep it has been noted that nerve pathways through the brain stem are unenergized. The head lolls limply. The body seems to be in a state half way to that known to forensic and other forms of medicine as 'brain death'.

So while the brain is really intensely active and engaged in processing, cross-referencing and filing vast quantities of data, the individual person is no longer being generated and hence no longer exists. It is, as we have already suggested, likely that the cyclic generation of consciousness which is continuous during waking hours is macroscopically reflected by the alpha wave produced by the waking brain. It is a fact that alpha wave production simply ceases during sleep. Consciousness has not, in the sleeping person, just 'gone underground'. It has gone completely.

It is a corollary of this fact that in true sleep the person cannot be dreaming. The activity that forms the basis for dreaming — the cross-referencing and filing of sensory data that is being downloaded from the short-term memory system — will be taking place but this is a process completely independent of the individual consciousness. It is not even precisely accurate to say that the person is still breathing. The truth is that the body, during sleep, comes under the control of endogenous life-support systems which are independent of consciousness. There is thus during sleep no 'person' present beneath the skin. No authority remains in the body that has any executive or decision-making capability.

It is worth noting that the endogeneous systems which keep the body running during sleep are not as sensitive or diagnostically efficient as those associated with the conscious state and for this reason the sleeping body is far more vulnerable to mishap than the waking one. A sleeping person is prone to death from causes, such as drowning in the bath, that could never happen to a waking individual.

Why is the mind switched off at night? There seem to be two chief possibilities. In the first place, it may be that for the downloading of diurnal sensory impressions into the archival memory, consciousness might simply be surplus to requirements. In this case, and taking into account the well-known tendency of all life-forms to conserve energy (as in hibernation) whenever possible, consciousness would be switched off simply because it would not be energy-economic to maintain.

Then again, it is well-known that sleep is not primarily designed as a rest period, but it remains the case that the body does need periodic rest and it may be that refreshing rest is hard to achieve during those periods when intensive neural information processing is taking place. The presence of consciousness during this period might make it very difficult for the body or mind to obtain even the relatively small degree of rest it requires.

More probably, the main reason why human beings, and other life forms, switch themselves off during sleep is that all or most of the consciousness-generating circuitry is needed for the immense alternative task of cross-referencing and filing the diurnal sensory impressions. It is likely that the neural circuitry simply cannot handle data streams moving both to and from the memory archive simultaneously or at least without the risk of dangerous interference and/or short-circuiting. Rather than risk malfunction it is far better simply to close down the mind-generating circuitry and concentrate all available electro-chemical resources on information processing and storage.

But there may also be more subtle reasons why the individual shuts him or herself down for the night. It may be that discontinuity of existence is indispensable for the full assimilation of the preceding period of conscious experience into ongoing personality evolution. It may also be that the shutdown permits refurbishment of damaged circuity (repair work). Research will be required to ascertain exactly which are the dominant causes. There is, however, no doubt that everything that we understand by the notion of a person vanishes when true sleep sets in. All bodily systems continue to function but now under automatic control. The mind is no longer being generated. The individual person no longer exists. He, or perhaps the pronoun 'it' would be more suitable in this context, has transformed itself into an immensely complex, automatic, electro-chemical data processing system that is operating 'at full capacity' without any personality participation.

The person, during sleep, can therefore no longer be considered to be 'animate'. For this reason it seems that it will be necessary for people to accept the psychologically and philosophically challenging notion that every single one of us becomes something conceptually very close to being a bionic entity, a 'cyborg', or, in the older SF jargon, a robot for one third of our lives on the planet. I myself find this notion disturbing.

What then is switched off? What exactly is the nature of the human mind which no longer exists during sleep? In terms of the Twin-Data-Stream Theory, a person can be defined as a self-conscious entity, possessed of some degree of autonomy, that is capable of meaningfully inhabiting space and time and of purposefully interacting with environmental events. The mind of such a person, as we have seen, can be regarded as the point of temporal confluence of two major and perhaps a number of minor streams of input data occurring within its brain. All these qualities apply to a waking human being but not to a sleeping one.

It is time to examine in some detail the structure and dynamics of the waking human mind. Unlike the sleeping mind, the waking one is endowed with a degree of coherence and enduring self-consciousness which survives the gaps in mind generation represented by sleep periods. Let us ask initially: where does the coherence and individual identity which help to define the concept of a person come from? My monitoring of my own mental processes suggests that the brain possesses and makes abundant use of a wide range of special-purpose 'drone' functions, all of which tend to fortify the mental coherence and well-being of the individual person.

The brain manifests, for example, a kind of 'need to keep in mind' function. I have frequently observed that, when I am concentrating hard on some problem, this function may repeat in the background of awareness a reminder of an urgent task. Thus while actually writing, I have sometimes been intermittently aware of a voice, which I have recognized as my own voice internalized, repeating monotonously a phrase such as: 'phone the doctor — phone the doctor — phone the doctor'.

In a similar way it is quite possible that in its conscious, diurnal mode the brain generates and sustains permanently a kind of basic data signal consisting of a person's name, age, and a few other vital facts. The chief purpose of this signal would be to provide a durable terminal around which the big data flows could perpetually cohere and organize consciousness. There are, as we have seen, two of these major data flows although they can be subdivided into tributory streams.

The first flow, which consists of incoming sense impressions, arrives as five sub-streams corresponding to the senses and also a flow of data produced by the brain's monitoring of endogenous processes. The chief function of the endogenous monitoring is doubtless to provide as early a warning as possible of bodily malfunctions revealed by pain, breathlessness and so on.

The second major flow is the oscillatory one stemming from archival memory. On moving towards consciousness this stream is filtered through a graded cascade of screens in order that data will ultimately reach consciousness strictly on a 'need to know' basis. The confluence of the two major streams at each experiential moment of existence is what we perceive as human consciousness.

It seems likely that the electro-chemical organization of the nervous system has been evolved for the purpose of isolating and hence rendering effective each brief (perhaps one-tenth of a second) reconstruction of the mind. Otherwise it is hard to see the purpose of conveying neural signals by the enormously elaborate method of continually switching the signals from chemical to electrical transmitters and back. But if the ultimate aim of this technique is the generation of a phasic constituent of mind then the technique would be explicable as providing a kind of transistor.

The synaptic chemical neuro-transmitters would, in their capacity as transistors, be able either to isolate or open up patterns of organized data in the brain. These fleeting patterns would form the physical expression of the perpetually modified individual mind. Every successive state (phase of consciousness) would thus consist of a neural network formed at that moment from amongst context-selected data cells scattered throughout the brain and probably throughout the entire nervous system.

In this connection, it is worth mentioning that there is probably no, or only very little, difference, as regards playing a functional part in the generation of consciousness, between neurons that are located centrally in the brain and those found in the periphery of the body. A memory may generate peripheral neural involvement. As a child, I crushed the fingers of one hand in some machinery. I have often noted that when this incident comes to mind I am aware of a faint reminiscent tingling in the long-since healed fingers of that hand. This may, of course, be explicable as a minor example of the 'phantom limb' phenomenon. But since signals pass continuously between peripheral nerve cells and cerebral ones it seems reasonable to suppose that structurally the brain can be regarded as permeating the entire body.

A human mind thus consists of the life-long sequence of the data patterns that form consciousness. Consciousness is modified perhaps ten times a second in phase with the alpha wave activity generated by the brain and probably the entire cerebro-neural system. The alpha wave is likely to represent the power surges needed for mind formation. The data patterns that manifest themselves as phases of consciousness result from the 'confluence' of simulated data flows stemming from sensory environmental monitoring and from archival memory. Each such fleeting manifestaion of 'mind' can be metaphorically imagined as a dome studded with millions of lights of which an apparently random, albeit high, proportion are lit. This pattern then, in a serial and coherent way, changes and develops at the rate of (perhaps) ten transformations a second.

The vast majority of the data cells from one phase of consciousness are reactivated when the next phase is formed a tenth of a second later. A hundred activations (ten seconds) later perhaps only a millionth of the same data cells will fail to have been reactivated while approximately the same number of previously inactivated cells will have achieved activation. The mind will thus contain a tiny amount of new material. Consciousness, in this scheme, is generated and regenerated six hundred times a minute. Possibly a half to two-thirds of all data cells are reactivated regularly throughout a lifetime and provide the substantial continuity of character and personality normally found in human beings.

The data flows are continuous but the generation of the mind is phasic, a process which is technically produced by the transistor-like switching potential of the chemical neuro-transmitters. It is thus clear that the long quest to locate the site in the brain at which the mind is located has always been a wild goose chase. The proper question to ask was never: 'where is the mind?' but 'when is the mind?' Spatially the mind consists of the sequential animation of a selection of data cells drawn from a pool that is coextensive with the entire human nervous system. But temporally it only exists at the present moment. It lasts for perhaps a tenth of a second and then the brain generates a 'new mind', or phase of consciousness, to replace it.

The conventional view of the mind has normally been some variation on the idea of a 'package' of reasonably durable 'thoughts' of which, at any given moment, some are conscious while most are unconscious. The basic concept here is that of an iceberg in which the greater part is concealed beneath the surface of the water. It is assumed that the mind is divided into two major portions, one of which is, as it were, above the surface of consciousness and the other below. But the mind as a whole is seen as a more or less static and enduring creation.

As we have seen, however, the true nature of the mind is dynamic and phasic. It is a perpetually self-transforming entity which can be conceived as an oscillating and gradually changing swirl of language-bound data particles (which might be called 'mentons'). There is no such thing as the 'subconscious mind'. The nearest mental constituent to it that really exists is the archival memory but this has little in common with the classical notion of the subconscious.

The mind as a whole can most accurately be regarded as the turbulent mixing of data particles formed temporally at the present moment (or a few milliseconds 'after' it) and derived phasically from two streams of information, one from sensory monitoring of the environment while the other is a complementary flow of relevant data from the archival memory which is perpetually summoned by the impact of the first flow. The generation of consciousness is phasic and each phase lasts a tenth of a second.

The movement of the brain through time generates the mind in a way analogous to that in which the movement of a magnet through a conductive coil generates an electric current. Without being passed continuously through what might be called an 'experience field' the brain cannot and does not generate mind. The sole continuous element of the mind — the archival memory — is itself in a perpetual state of modification and hence evolution.

Moreoever, the archival memory has no direct interface with consciousness but only exists functionally as the transponder-like source of one of the two main data streams that compose mind. This is the reaon why it is so hard for a person to remember simply by willing to do so and also the reason why an individual may be tormented by undesired memories.

The fact that the mind is composed of a data flow from incoming sense impressions meeting a data flow from archival memory means that 'consciousness' is several milliseconds behind the 'true' present. The need for a reflex action system to cope with sudden emergencies (say a car abruptly emerging from a side street) is probably a result of this time-lag. The delay may also be responsible for certain curious facts associated with surgical experiments on direct stimulation of the brain that Daniel C. Dennet, amongst others, describes (notably in his work *Consciousness Explained*). In these experiments, patients, when the appropriate cortical nerve cells have been methodically stimulated in an experimental set-up, have seemed to detect an effect that precedes its cause.

It is the phasic nature of mind which may be responsible for the strange results of certain other experiments. In them light signals appear to change colour 'in flight' and words disappear from, and appear upon, a computer screen at the point of arrival of the trajectory of the eyeball of a person scanning that screen. The subject does not himself perceive any change in the words on the screen ahead of him. This fact is inexplicable according to classical theories of the brain/mind complex. But the Twin-Data-Stream Theory explains it by stating that each change of wording on the screen occurs during the interval between successive, ten-times-per-second mind activations. This results from the fact that the word changes are computer coordinated with the rate of saccade (rapid movement) of the eyeball.

In fact, mind generation is a kind of interior strobing. Just as the exterior strobing produced by a film projector can seem to halt, or even reverse, the apparent turning of wagon wheels in a movie, so mind generation strobing can generate apparent rotary motion in a still image.

As further potential confirmation of the phasic generation of consciousness is the strong possibility that the phenomenon of flicker epilepsy — a kind of epiletic fit generated in susceptible subjects by a rapidly flickering light — might be caused by an interference pattern set up between the flickering light source and the frequency of the generation of the mind. Indeed, once it is recognized that the mind is not continuous but is generated at a particular frequency then our understanding of a wide range of phenomena, from mathematical and musical awareness and ability to athletic and other skills (and also, of course, deficits) will doubtless be enhanced.

A sleeping person has virtually ceased to inhabit time. A waking one is almost infinitely time-responsive. It is temporal variation which directly generates the master stream of consciousness, i.e. the data stream stemming from sensory environmental monitoring. This master stream in turn calls up the tributory stream of data from archival memory automatically. The turbulence of the confluence of the two streams constitutes human consciousness.

As a consequence of this mechanism our minds are so exquisitely responsive to, and dependent upon, temporal variation that they are only capable of survival when immersed in the swift-runnng river of time. If either of the two major data streams slows down and stops then our minds dim and finally go out as a consequence of the deficit. But much of this is as true of an animal mind as it is of a human one. For this reason, we must now look at the crucially important mental constituent which divides us and our minds sharply from those of the rest of the animal kingdom: language.

We have seen that from birth onwards each individual starts to amass an archival memory. The progressive stocking of the archive continues in a similar way in humans and animals (baby chimps outperform human babies) until a crucial stage is reached: the acquisition of language. With this new and overwhelmingly significant acquisition, life that is essentially human can be said to begin. Certainly the superb phrase which opens St John's Gospel — 'In the beginning was the word' — applies just as much to the evolution of each individual mind as it does to the greater sphere for which it was originally designed; the universe itself.

As knowledge of, and fluency (the ability to generate a flow) in the use of, language become increasingly comprehensive, human consciousness becomes increasingly bound by the fusion of language and data into a mind that is evolving towards self-awareness. It is this self-awareness or language-mediated consciousness which, wholly and exclusively, is the great differentiator between animal and human minds.

Animals experience REM sleep and store sense impressions for help in navigation, food finding, distinguishing useful elements in the environment from useless ones and friendly organisms from hostile ones. Non-human animals, in exactly the same way as humans, have minds. The possession or lack of a mind does not define the difference between human existence and existence as experienced by other animals. If anything, it is more likely to define, or at least crucially affect the definition of, the difference between living and dead matter.

A mind, as we have seen, is simply the confluence of two data streams, one deriving from present and one from past experience. All living organisms — even bacteria or viruses — must have such a mechanism, however rudimentary it may be, in order to live. It was probably the fortuitous acquisition by certain kinds of inert matter (clays have been suggested) of this ability which generated the original living organisms on earth.

Nevertheless, animals lacking language cannot, however large their brains, distinguish between 'I' and 'not-I'. This is the initial and crucial distinction that ultimately makes possible the concept of an objective universe which can be consciously inhabited and purposefully manipulated. Minds informed by language can then go on to make the next crucial distinction which is that between time and space. From there on language can be used to analyze the universe into ever more detailed constituent parts. Organisms without language cannot take even the first step in this march towards self-consciousness and domination of the environment

Zoologists have known for some time that certain species of dolphin have brains that, both in absolute terms and relative to body size, are larger than our own. *Tursiops truncatus*, the bottle-nosed dolphin, and other species of dolphin, are also endowed with a cerebral cortex that is more convoluted than ours (meaning that if 'spread out' it would provide a larger surface and hence signify at least potentially higher intelligence).

In other words from every macroscopic perspective some dolphins should oustrip us mentally. And yet they have not, as far as can be determined, taken even the first step towards achieving civilization. There may be physiological or environmental or even acoustic contributory factors inhibiting dolphins from producing a culture. However, the immediate reason why dolphins have not made any significant move in this direction is that they have not evolved a syntactical, verbal language in which they can separate 'I' from 'not-I' and so start 'talking to themselves'.

At the most fundamental level, it is the ability to 'talk to ourselves' which makes us human. Mention was made, earlier in this book, of the fact that the brain manifests various semi-automatic facilities such as an urgent memo signal — 'call the doctor — call the doctor' — and a personal data signal for continually transmitting an identity check into consciousness. But the chief of the brain's drone functions is what might be called the verbal commentary.

If we watch infants who have recently acquired language we can often both see and hear them experimenting with the production of this verbal commentary which consists of telling themselves the 'story of me'. A little girl may say to her doll: 'Now I have to dress you and put you to bed and then — and then — I have to wash you and — no, first I have to wash you and then put you to bed and then — oh, here is the sweet I was sucking and its all dusty — I wonder if I could still eat it?'

This ceaseless, but perhaps intermittent, self-commentary continues until death although it can become so sublimated and veiled by foreground consciousness in an adult — and especially a linguistically sophisticated adult — that he or she may, for long stretches of time, completely lose awareness of the fact that it is taking place. Nevertheless, it is this life commentary, told by the first person to the first person, which, more than any other product of the human brain, comes closest to the essence of being human.

It is clear that language and mind are so inextricably intertwined in a human being that each is impossible without the other. This makes it a question of immense interest, although perhaps the answer lies beyond our grasp, as to how we acquired language in the first place. Increasing brain size is the *sine qua non* of ultimate language generation but it is not in itself, as we can discern from the situation of the very large-brained, non-human mammals, sufficient for the task. So how did big brains come to be selected for human beings by the mechanism of evolution? The following passage is an attempt at providing a schematic answer.

The growth of the human brain to a size potentially capable of deploying language may have been the result of a serious, and perhaps near-fatal, environmental dilemma confronting early hominids. This potentially disastrous situation would have resulted from the fact that the hominids had abandoned the trees for a bipedal existence on the savannah. However, they were frail creatures who were, in most physical respects, very ill-fitted for the challenges of the new environment. Indeed hominids having the physical characteristics of the three-million-year-old australopithicine specimen known as 'Lucy' would have been amongst the most vulnerable of mammalian savannah dwellers.

They were small, weak, slow and largely defenceless. They had very little physical prowess but they did have one potentially valuable quality. They were relatively intelligent. Since nothing much could be achieved by evolution in the direction of improving their spindly physical structure, the Darwinian sifting process concentrated on intelligence. The hominid brain began to grow dramatically. It turned out, however, that a huge amount of additional brain tissue was needed for quite modest gains in survival potential. Even stone tools, once the new intelligence had made these possible, did not make our frail forebears a match for the formidable power of the predators.

Nevertheless, for want of any better procedure, evolution (which, the author concedes, is being shamelessly anthropomorphized for this episode) simply continued to increase the size of the hominid brain. And finally this strategy of despair paid off. Our brains attained the critical degree of size and power needed for the generation of syntactical, verbal language. This epochal development utterly transformed the evolutionary situation of the hominids.

Language made possible organized and complex cooperation between individuals. As soon as this happened the hominids were transformed from being little more than a tasty morsel for a leopard into the functional likeness of a new, super-organism. They became capable of cooperative action involving the whole family or even tribe. Anyone who has seen, probably on a television screen, the clumsy hunting techniques employed by pack hunters such as wolves or even chimpanzees will realize that huge increases in efficiency would be achieved if the animals could plan both strategy and tactics in advance of the hunt and also communicate detailed messages throughout it.

Having achieved language the new 'talking hominids' would have become capable of such verbally-mediated cooperation. As a result they would have not only been able to immensely improve their food supply but also to neutralize at least the worst of the threat posed by the slow-witted predators. From precariously situated survivors they would rapidly have graduated to being lords of the savannah.

And they would also, by the time they had evolved this far, have started telling themselves the 'story of me'.

## The 'Story of Me'

The universe, from a human perspective, seems to be fundamentally discontinuous. There are large distances between planets, far greater ones between stars and almost inconceivable ones between galaxies. At the level of the very small, it seems that however much physicists and mathematicians divide and sub-divide matter they never reach the level of the indivisible. At least, they have not done so to date and are currently speculating about the possibility of units of matter, known as 'superstrings', which bear the same ratio to the size of atoms as atoms do to the size of the solar system.

The earth and its surface-dwellers form something like a continuum but the substance of which our planet and its inhabitants is composed can itself be divided into fairly discrete chemical elements. Life can be conceptually separated into distinct groups such as species, genus, class and so on. At the level of individuals, each living organism is a separate entity with more or less secure boundaries. And in terms of the theory of the structure and dynamics of the mind that has been proposed in this book these 'gaps in nature' extend to consciousness itself which is generated at a specific frequency and suspended at regular intervals throughout life. Thus, the universe appears to be, in whichever direction we look, formed from relatively discrete constituents.

Might there be, as a consequence, a primal tendency for an intelligent being to try and bridge, if only conceptually, these gaps? Is there a profound need for mind to see itself as part of a unity? If this were the case then a mental creation that might help to provide at least the illusion of a unity or continuum would be the story. A story, by means of its fundamental nature, bridges gaps. It can link far and near, then and now, 'me' and others. Ultimately perhaps a super-story could be composed which would bind the entire universe into a whole. Many of the world's religions can, in fact, be seen as attempts to do just this.

One of the profound 'thoughts' of the seventeenth-century French philosopher and mathematician, Blaise Pascal, is: 'the eternal silence of the infinite spaces frightens me'. This admission seems to suggest that the universe struck Pascal, a scientist, as being alien, meaninglessly (from a human perspective) vast and overwhelmingly hostile to life. But human beings have a powerful need to feel that they inhabit a secure and benign home which has at least some definable boundaries.

We are also aware that, as organisms that have themselves had a beginning, we tend to perceive all phenomena as having, like ourselves, a beginning, a middle and an end. The concepts of eternity and infinity are inherently vertiginous to us. Might part of the solution that we have found to the implied challenge of living in a universe that can only be defined by dimensions incommensurable with those of our own existence be the story?

A story permits us to pretend that eternity and infinity are not inherently incomprehensible but are at least cousins to the types of time and space within which we pass our lives. For this reason we talk and write as if there were nothing incongruous about conferring upon what is actually an arbitrary blink in the endless span of eternity (of which a single second is the same proportion as a million billion years) a specific name such as 'The Twentieth Century'. We even cheerfully subdivide such cosmic blinks into cosy sub-blinks like 'the naughty nineties' or 'the hungry thirties'.

Novels, plays, operas, gossip, journalism, narrative of all and any kind help bind the universe into at least the semblance of a whole and give us at least the illusion that we may ultimately find a way to domesticate it. There is a phrase which has been traditionally used in English to initiate the telling of a folk tale or legend. It is 'Once upon a time —'. There is, of course, actually no such thing as 'a time' — a discrete portion of eternity. But to help us bear the impersonal and unending flow of change we invent stories which recount how 'once upon a time' a handsome prince married a beautiful princess and they 'lived happily ever after'.

By means of such tales we at least seem to have redeemed a portion of eternity. We have, if only momentarily, breached the 'eternal silence of the infinite spaces' and filled it with human voices and human values. Our stories are not, perhaps, much of a victory over the vast impersonality of the cosmos but they are a great deal better than nothing. They may indeed be indispensable for maintaining the sanity of an intelligent species that has evolved language.

We can now attempt to relate these ideas to the subjects we have been considering. As regards dreams, the need to tell a story certainly seems to be a part of the impulse behind their generation. What returning consciousness actually apprehends, when it begins to awaken after a night's sleep, is a sequence of disconnected images, or perhaps images connected in a loose thematic way, but the natural imperative of the mind towards story-telling encourages consciousness to use them for the construction of a surreal tale.

The imperiousness of this urge to make stories can be further illuminated by looking at an account given by the neurologist, Oliver Sacks, of his experiences with patients suffering from the fearful affliction known as Korsakov's syndrome. The defining characteristic of this cerebral disorder is that the patient's capacity for remembering any new experience is reduced to a time span of a few minutes. A Korsakov sufferer can meet the same person several times an hour and each time be convinced that he or she does not know the other. The mind becomes a kind of 'memory sump' into which all new sense impressions swiftly sink and vanish. In terms of the Twin-Data-Stream Theory, it seems highly likely that the short-term memory is very largely disabled.

But failure of the short term memory is not all that the Korsakov sufferer has to contend with. In addition, the patient lacks any stocks of archival memory that refer to the period after the patient's brain was damaged by a viral disease or extreme alcoholic excess. It is worth pointing out that the fact of the correlation of the loss of the short term memory with the truncated condition of the archival memory is itself powerful testimony to the validity of the Twin-Data-Stream Theory. When the incoming sensory stream of data ceased to reach the short term memory then the archival memory ceased to be stocked.

One of the consequences of this appalling double deficit of memory is that sense impressions reaching the patient can summon up from archival memory little or nothing of much relevance to his present situation. The patient is fluent verbally since his archival memory stocks persist from before the traumatic event that prevented its further development and are replete with language-bound data. This data, however, derives from the patient's near or even distant past. Ever since that time the patient has inhabited a 'data desert'. The solution that such patients normally adopt to their desperate need for a data flow from archival memory relevant to their present situation is to cobble together a semi-fictional substitute for it. The patient tells himself an opportunistic and largely absurd 'story of me'.

The surreal fiction that results from the patient's frantic efforts necessarily derive exclusively from the stream of data reaching him from sensory monitoring of the environment. Sacks quotes some of the febrile monologues that are produced and they are revealed as being clearly analogous to dreams. This is because the etiology of the two phenomena is similar: both dreams and Korsakov narratives are stories that strive valiantly, but ultimately vainly, to bind incompatible items of information together into some kind of coherence.

Here is one of Sacks' Korsakov patients trying to fashion a story from intractable fragments of perception. This patient was, before suffering his cerebral lesion, a butcher. In the following fragment of dialogue he is talking to, or rather at, Sacks who is wearing a stethoscope around his neck. Although he sees Sacks daily and sometimes hourly, he has no notion of who the neurologist is although at the end of the fragment of monologue the white coat and stethoscope apparently provide him with a usable clue. Initially, however, he identifies Sacks with a former regular customer whom he can remember from the time before the memory stores in his brain were destroyed.

'You mechanics are all starting to fancy yourselves as doctors, what with your white coats and stethoscopes — as if you need a stethoscope to listen to a car! So, you're my old friend Manners from the Mobil station up the block, come in to get your boloney-on-rye . . . Where am I? I thought I was in my shop, doctor. My mind must have wandered . . .'

It is worth pointing out that the problem of Korsakov sufferers can be conceived as the opposite to the one (which we will discuss in a later chapter) that afflicts schizophrenics. The Korsakov patients have no, or very little, input from archival memory and schizophrenics have too much. The solutions that both kinds of patient tend to adopt are again opposed. The Korsakov victim employs frenzied generation of narrative in a hopeless attempt to bridge the appalling chasms in his memory. The schizophrenic attempts to reduce the archival memory input.

Sacks' own description of the efforts of the Korsakov sufferer to manufacture the 'story of me' is so good that I will relay a few words of it. In the following passage Sacks is commenting on the plight of the patient that I have already quoted above: 'He remembered nothing for more than a few seconds . . . Abysses of amnesia continually opened beneath him but he would bridge them, nimbly, by fluent confabulations and fictions of all kinds . . . (he) continually improvised a world around him — an Arabian Nights world, a phantasmagoria, a dream . . .'

The tragic situation of both Korsakov sufferer and schizophrenic point us in the direction of the same crucial question: why is the narrative impulse so overwhelmingly urgent in human beings? And the fundamental answer is that story-telling lies at the very heart of being human. The truth is that the human mind, which is a fusion of language with sense-impression and which distinguishes human beings from any other life-form of which we have any knowledge, is itself a story. More accurately it can be conceived as the product of two interactive types of story.

The first of these is fleeting and lasts a tenth of a second. In the course of a lifetime, we tell ourselves a huge number of these brief yarns. Indeed, by the time we have achieved our biblically allotted span of seventy years, each of us is the author of some sixteen billion of these tiny, complex and poignant tales. The second kind of story that we tell ourselves is much longer. It lasts a lifetime or at least that part of it which stretches from the time we acquire language until, usually in death, we lose it again.

The first of these two kinds of story is, of course, each individual phase of consciousness. Many readers will doubt that a phase of consciousness can be regarded as a story at all since it only lasts a tenth of a second and consists of nothing more than a brief swirl of data particles. It is a surprising fact, however, that these phases, while far from representing the conventional notion of a story nevertheless include all but one of the traditional components of linguistic narrative.

These components are: incident, character, description, emotion, sensory perception. All of these qualities are, of course, dependent on the manipulation of language for their realization. The only important characteristic of a conventional story which a phase of consciousness does not possess is a linear structure and it is precisely this lack that may make it hard for many people to accept that it can be a true story. A non-linear story, in which the elements interact without apparent causality or succession, will strike many people as a strange concept although it is far from unknown in modern literature.

Although lacking a linear structure, a phase of consciousness does, nevertheless, have a well known organizing principle. This is that of the field or constellation. For this reason a phase of consciouness can be thought of as a story in which the story-telling principle consists of narrative elements held in a state of dynamic suspension rather than unfolding sequentially. The 'story', suspended in each phase of consciousness, consists of sense data fused with language.

In terms of the structure of the human mind the field or constellation story does, in fact, reveal a potential to generate a linear narrative but only when these tiny tales are strung together to form a sequence. As this linkage proceeds (perhaps at the rate of ten manifestations per second) then the individual 'field stories' become components in the linear, time-hugging and lifelong 'story of me' which is, in fact, the human mind.

Because it consists of a fusion of sensory data with language human consciousness differs fundamentally from animal consciousness. It is, however, virtually impossible to discuss this difference without distorting the underlying reality. This is because discussion and analysis can only be performed with language, and language itself reflects the structure of the human mind. The subject/verb distinction, crucial to all human experience of the universe, has no place in animal existence. Therefore an animal does not 'do' anything at all. It becomes. When a dog runs, it essentially becomes the act of running. A wounded man will suffer but a wounded animal 'becomes suffering'. With this important caveat, then, let us make some attempt to show that while a human and an animal may both perceive the same aspects of the environment what they will *experience* will be completely different.

Let us imagine that a man and an animal both see a tree, hear a bell and smell burning. The man will receive each of these sense impressions with a linguistic tag attached to it saying 'tree' 'bell' 'fire'. There are no tags or markers that could help the animal understand these phenomena even if there could be, without language, an 'I' to do the understanding. For this reason, and no matter how large the animal's brain, it is unable to separate conceptually the tree from the field in which it grows, the bell from any other background sound, the smell of burning (other than by means of a fear reflex) from that of dung or hay or itself.

So the animal, lacking language, cannot tell itself a story enabling it to grasp the significance of the sensory data it has received and also, if necessary, to take appropriate action. Should the animal be harmed by the fire that is generating the smell of burning then, upon subsequently perceiving the same smell, it might react with reflex fear. But it can never say to itself, as humans can and do: 'I smell burning. I wonder if something's on fire? Perhaps that was a warning bell. Let's see, the grass in the meadow is very dry. The tree might catch. I could be trapped by the flames and so I'd better be careful how I approach.'

Animal consciousness is composed exclusively of particles of sensory data, whether directly apprehended or received reflexively from archival memory. Human consciousness is made up of cognate particles but fused with language. It is the language component which successively organizes itself into a continuous narrative and that narrative is in fact the 'story of me'. And the 'story of me' is the human mind. Animals, unable to differentiate themselves from any other phenomena, do not generate a 'story of me' and thus, in the human sense, do not have a mind but only an endless sequence of phases of consciousness.

It might be helpful if we named the language-bound data particles, which are exclusively the product of a human brain, 'mentons' to distinguish them from the animal equivalent which could be dubbed, not very euphoniously, 'sensons'. A 'senson', in this scheme, is purely a particle of stored sensory impression. A 'menton' is a similar particle but one which has, during the process of consciousness generation, become fused with, and therefore tagged by, language.

The pre-human biological mind — the mind of mammals and also of lower forms of life probably down to the bacterial or even viral level — is, like the human mind, generated by the confluence of twin data streams. The first of these streams is composed of sense impressions from the external world and the second of items automatically evoked from archival memory by the impact of the first stream. But the minds of animals can only hold raw data which has no language component. Animals therefore have no capacity for interpretation, combination, elaboration or any conceptual manipulation of their environment.

This limitation is, to give a striking example, the reason why a giraffe, in a famous series of photographs, is shown dipping its head into a river to drink immediately after having beheld the raised head of a neighbouring giraffe, which had clearly a moment before been drinking, clamped in the jaws of a dangling crocodile. Lacking the ability to interpret the significance of the scene by the use of language, the thus far unharmed giraffe could only have perceived its neighbour as 'a thing' with a long scaly snout. What a human being, informed by the linguistic tags attached to the visual data, would have seen was a 'CROCODILE' attacking a 'GIRAFFE'.

Even an animal with a brain as large as a human being's would have been unable, lacking language, to interpret the scene with any more depth of understanding than the observing giraffe could do. Giraffes are not, of course, especially intelligent but it was not sheer stupidity which governed the observing giraffe's apparently suicidal behaviour. It was the lack of language with which to analyze the situation and thus reveal to itself that the stream into which it was about to dip its head contained mortal danger.

But, of course, language enables a human mind to do vastly more than distinguish between GIRAFFES and CROCODILES even when one of the latter is spatially attached to one of the former and thus produces a continuous, if novel, visual image. Language enables us to understand and manipulate the entire universe. Language may, in this context, be conceived as a kind of magic knife with which we can cut up the universe into manipulable sections which we can then rearrange according to need or even whim.

Learning from experience — other than by the mechanical operation of the conditioned reflex — is virtually impossible without language. In the same way, all the more abstract and combinatory structures of the mind are crucially dependent on language. The basic truth is that without language there can be no true mind and no extension of the notion of experience in the direction of either the past or the future.

Animals are therefore locked into the passing moment and can never develop self-consciousness since the key element in the construction of self-consciousness is the fusion of data with language. Some dolphins have bigger brains than we do. This implies that they have, at least potentially, a more efficient data processing capacity than we do. Nevertheless, lacking language they remain unable to generate self-consciousness and thus master their environment.

What are some of the consequences of lacking, and also of possessing, verbal language? Animals that lack it are not born and nor do they die — except to a human observer. The animal, having no knowledge of either birth or death, is in a sense eternal. In another sense it is instantaneous since it can never escape from the passing moment. There is no past and no future for any being lacking language. Animals are spared the terror of the approach of death but they also lack the capacity to delight in being alive. We humans, unlike animals, may experience loneliness but we also know love and friendship.

Animals, lacking the ability to tell themselves the 'story of me', cannot distinguish conceptually between themselves and others. They can make this distinction functionally as a result of instinct but they have no knowledge of its implications. Again, human beings, unlike animals, can generate culture and civilization but we also experience the inevitable negative consequence of these structures which are things like war and hatred. Without language, animals, no matter how large their brains, cannot practice science, art or mathematics. They are thus categorically barred from influencing the conditions of their own existence or their further evolution. Beyond the behavioural characteristics and instinctual imperatives lodged in their genes, they are purely creatures of chance and hazard.

Clearly the gulf between animal consciousness, composed exclusively of raw data without interpretation, and human self-consciousness, is enormous. However, a remarkable hint (no more) of its nature is contained in some of the writings of Helen Keller, the American authoress of the early part of this century who became totally blind and deaf at the age of nineteen months. The extraordinary efforts of dedicated teachers ultimately enabled her to learn to communicate first by sign language and finally by speech.

She has written: 'Before my teacher came to me, I did not know that I am. I lived in a world that was a no-world. I cannot hope to describe adequately that unconscious, yet conscious, time of nothingness ... Since I had no power of thought, I did not compare one mental state with another.'

This is clearly a valiant but doomed attempt to verbalize the condition of being non-verbal and gives a clue, although the variations in the two states are still huge, to the difference between existence as experienced by even a large-brained animal without language and a human being in possession of a language-based mind capable of generating self-consciousness.

There has been much discussion in recent years as to whether language is or is not indispensable for thought. This speculation is rather surprising since the question can be settled conclusively in under a minute by anyone prepared to perform a simple subjective experiment. This takes the form of attempting to 'think' for, say, thirty seconds without making use of language. He or she will find that it is impossible to do so.

Some may claim, having attempted the experiment, that they successfully avoided the use of words but still achieved 'thought' by generating a meaningful sequence of images. Even if we concede that a sequence of images constitutes a functionally valid form of thought a little reflection (necessarily using language) will immediately show that, without some degree of verbal ordering, the images would have been purely chaotic or directed towards the satisfaction of a basic instinctual goal. No sequence of images, undirected by language, can be relevant to the contemplation of a complex human issue or the solution of a human problem.

Language, as we have earlier shown, is the indispensable instrument for breaking down the environment into components that are either mentally or physically (or both) manipulable. For this reason, any possible sequence of images in a human mind will necessarily have itself been ordered, just as it must have been summoned, by language. Without language not only is it impossible to break down the sensory field into usable parts but it is even more fundamentally impossible to separate the perceiving individual from the perceived reality. Without language there can be no 'I' and no 'it' and, as a consequence, no 'objective reality'.

The primal distinction between what is and what is not 'the self' is the one that a human baby makes with its first meaning-laden utterance. When it first says 'ma' (most commonly) or its equivalent in some other language, and thus attaches a repeatable linguistic tag to an aspect of external reality, it makes the crucial separation (which even the largest-brained dolphin or elephant cannot make) between that which is itself and that which is the rest of the cosmos.

Animals therefore do not, and cannot, have any personal identity. They are unable to distinguish between themselves and their environment. More fundamentally they are unable even to form the aspiration to do so. In fact, in a philosophically pertinent sense, it might be said that animals do not actually 'live' at all although they are, of course, alive. This is because the verb 'to live', as we apply it to our own lives, implies two interactive elements: the being that does the living and the life that is lived. But this is a distinction, like all the others which define the difference between a human and an animal mind, that can only be made with language.

Without language there will, of course, still be vital process but there will be no knowledge of it. Animals are themselves life but only human beings actively 'live' in the sense that their thought, originating in experience which may be more or less the same as animal experience, can potentially expand outwards to embrace every real and also imaginary aspect of the universe. Animals, lacking language, can never be more than co-extensive with their own sensory impressions and genetic imperatives.

It might be argued that language is not essential to thought since a mathematician can think in numbers or a musician in notes but this is a superficial objection. The notes and the numbers constitute a special form of language, a linguistic shorthand. Everything that can be said in either of them, or in other codes (which incidentally can only be learned by the use of language), can also be expressed using words.

Despite these considerations, certain philosophers, such as Bryan Magee in his entertaining book *Confessions of a Philosopher*, argue passionately that 'the most important . . . of our dispositions cannot be put into words — being in love, for example, or our feelings for our friends, or what music means to us in our lives, or our relationship with our children, or our passion for philsosophy . . . none, or almost none . . . can be adequately expressed in language.'

This assertion neatly expresses the precise opposite of the truth. In fact, it is only and exclusively through the use of language that the matters Magee lists can be given expression or even (other than in terms of faint, inchoate stirrings) felt by an intelligent being. No matter how great the brain size or potential intelligence of any non-human animal, that animal cannot, without language, comprehend the simplest of Magee's notions — nor even the more fundamental fact that it exists at all.

If it were possible to erase all trace of language from a person then he or she would be instantly reduced to the level of an animal. 'He' would have no knowledge that 'he' existed or would one day cease to exist. Nor would 'he' know, deprived of all objective knowledge, that 'he' had physical boundaries or that the world had both spatial and temporal dimensions. In the sense that human beings, who cannot think without using language, 'know' things, animals are totally incapable of knowing anything.

So how is it possible for philosophers like Magee to credit animals with thought processes analogous to human ones? The clue may be found in a later passage in Magee's book where he writes: 'A lot of our connected behaviour is similar to that of animals but no-one supposes that animals think in language'. This is true but represents rather sloppy thinking for a philosopher. A simple syllogism is enough to reveal the fallacy involved. This is the belief that because two entities have elements in common they necessarily share other components too.

Magee is, of course, right in thinking that we share a lot of our behaviour patterns with animals and the reason for this is simple. We really are animals. But we are not only animals. The possession of language makes us both different from and, in one sense at least, superior to non-verbal animals. In addition to the animal substrate, we are also linguistic entities and, as such, we may have, and often do have, different horizons and purposes from animals.

The important structural distinction between the consciousness of an animal and that of a human derives from the fact that our memories store not only raw sense data, like those of animals, but also the interpreted, combinatory and coded data we call language. And language is not just an added quantity, a utilitarian supplement to enhance the reasoning powers of a horse or dog or a whale (none of which, in the human sense, actually has any reasoning power) but represents what is actually a different and evolved being that shares the body of the host animal.

It is probably advisable here to confront an implication that has been hovering uneasily about this chapter. It is an implication that may be partly responsible for the reluctance of thinkers like Bryan Magee to acknowledge the crucial role played by language in any useful definition of what it means to be human. This is the notion that the condition of a human being somewhat resembles that of a parasite exploiting an animal's body.

In fact, from a pejorative point of view, this is a valid way to describe our nature and our relationship with the animal world. Its possibility has been intuitively perceived and expressed throughout the civilized ages in notions such as that of the twin natures of man, part God and part beast, the Christian concept of the body being merely the container of the 'immortal soul' or the philosophical idea of the 'ghost in the machine'.

In demonstrable fact, we really are to some extent parasitic on the big-brained primate beneath our mental skins. We have usurped its biological imperatives to feed our own bloodlessly cerebral ones. We have acquired, through the use of language, values, horizons, concepts, behaviour patterns, social structures and most significantly our very beings — all of which are alien to, and quite often destructive of, the quality of a purely animal existence.

In place of the careless delight (as we interpret it) of an animal in unmodified nature — a swallow swooping through the sky or a falcon tearing at a piece of carrion — we have created a universe of guilt and innocence, of pleasure and pain, of knowledge and of ignorance, of life and of death, of dirt and cleanliness, of dimension in time and in space which are all foreign to the purposes of non-linguistic life on our planet.

Many of the things an animal does naturally and without shame — copulation, excretion — we do uneasily, aware that such biological acts are remote from the value systems we have generated and that they awaken in us inescapable and testing moral quandaries. The things that we delight in and achieve naturally — friendship, knowledge, comfort, art, science and also alas organized destructive activities — have no counterpart in an animal's existence.

So it seems hard to deny that, from an animal point of view, we really are a kind of parasite. But that very declaration points to an escape route from such an abject perspective. Why? Because animals do not have points of view. A point of view is purely language-generated and hence human. So we can boldly proclaim that from our human point of view we are not parasites but a breed of super-animals. We are beings that combine the physical nature and vitality of animals with a mental freedom and inherent questing drive that carries us soaring above an animal existence into the possibility at least of alliance with the fundamental processes of the universe.

Within an evolutionary time-scale, intelligent language-oriented human beings have lived for only an instant. It is thus far too early to expect that the relationship between our human and animal natures has had time to become harmonious and mutually supportive. For many thousands, perhaps millions, of years to come, this relationship will be testing and difficult. A human being is an animal that has generated within its nervous system a semi-autonomous linguistic being, a genuine 'ghost in the machine', which inhabits the same material body but only partially and occasionally shares the same imperatives.

Support for this idea can be found by looking at a phenomenon that has always puzzled scientists and ordinary people alike. It is called infantile amnesia. Why do human beings invariably forget, or seem to forget, everything about their first year or two of life? The brain of a baby is known to be more active and acquisitive than it will ever be again. In later life it is quite easy to remember events that may have occurred half a century before. Why then should a human being by the age of, say, five have totally forgotten what it was like to be one?

The answer is that the five-year-old has become a radically different being from the one-year-old. It is not that he or she has forgotten what it was like to be one but that he or she never knew. The one-year-old was a pre-linguistic animal rather than a human being and therefore knew nothing. When a human being remembers its past self it remembers episodes of 'the story of me', i.e. sensory experience tagged with language. But the one-year-old, although its archival memory will have been immensely active (which is paradoxically why babies sleep so much) in storing sense impressions, will not as yet have been able to tag these impressions with language.

This means that there will simply be nothing verbal, and hence reproducible, in the mind (that is the archival memory) of the five-year-old which remains from the earlier 'pre-verbal' stage of its existence. It is not that it has forgotten what it was like to be an animal. It is rather that, as with all the other animals in the world, such knowledge does not exist. No animal 'knows' what it is like to be an animal because to 'know' anything language is indispensable.

As the baby progressively accumulates a linguistic identity it becomes more and more human. What is inaccurately called 'infantile amnesia', and is really simply ignorance of the mental processes of an alien life-form (an animal), is progressively dissipated. By the time the child has reached three or four, and almost certainly by the time it is five, it has become much more human (i.e. a creature of language) than it is animal and so its archival memory has become capable of stocking human memories.

In other words, none of us (except for a few extremely precocious language achievers) ever has been a one-year-old human-being. There is no such thing. We were only one-year-old animals which required another year or two in order to 'secrete' the infant linguistic entity which becomes the human being. Since the new human being has a qualitatively different mind from the initial animal there will be no, or hardly any, interpretable memory traces of the early period stored in its archival memory.

We are creatures of language. Storage of experience is a process common to animals and human beings alike. But whereas we share basic sensory consciousness with animals it is language that makes possible that immensely important attribute which no animal other than the human animal possesses: 'self-consciousness'. It is self-consciousness which makes us human. It will be by examining the relationship between language and experience more profoundly that we should be able, in the next chapter, to achieve an understanding of the astonishing nature of the human mind itself.

# Beings

The term 'being' suggests, if only because of its abstraction, something possessing few if any tangible qualities but possibly endowed with considerable potential. Over the millennia, our species has given itself many names ranging from the prosaic 'mankind' to the coldly classificatory 'homo sapiens'. But the one which, judging from the mildly approving way in which we normally use it, best expresses our own sense of ourselves is 'human beings'.

In this context, it is instructive to recall that the Old Testament credits God with the self-definition: 'I am that am'. This claim seems to suggest that deity (at least in the estimate of the profound thinkers who wrote the Judeo-Christian core texts) conceives of itself as being composed of pure 'existence'. Using the same metaphorical basis, it is clear that an animal — sheep, dog, bear, *staphylococcus aureus* — is not a 'being' at all but merely an aspect of the universe to which a language-endowed mind — our own — has attached a convenient label. Animals cannot, like deity and, in due degree, like ourselves, be meaningfully thought of as being 'beings'.

Or can they? On a country walk some years ago, while observing a small flock of sheep, I asked myself if these familiar denizens of the English landscape might seem a little less prosaic if I tried to think of them as 'ovine beings'. This minor linguistic adjustment had an immediate effect. Around me still grazed the same black-snouted, woolly quadrupeds but now each animal seemed to have acquired a new dignity. Thought of as 'ovine beings' my four-legged companions had gained at least a faint aura of mystery and potency. To put it a shade barbarously, while a sheep may strike a human being (vegetarians excepted) as little more than meat on the hoof an 'ovine being' becomes, potentially at least, a brother.

But, of course, as far as the individuals that composed the grazing flock were concerned they not only failed to qualify as ovine beings but could not even achieve the relatively humble status of sheep. In fact, the creatures were (to themselves) nothing at all for they had no language to generate the self-consciousness necessary to know that they were discrete entities or even conceptually to distinguish themselves from their physical environment.

We have seen from earlier chapters that a human being is actually a verbal entity continually assembled and reassembled (probably ten times a second) in the neural tissue of what, without its presence, would be merely an animal, perhaps a higher primate similar to an ape. I must now backtrack somewhat to stress that this notion should not be seen as implying that a human being's mind is materially different from the mind of an animal. The repeatedly generated person uses the electro-chemical pathways of the nervous system to manipulate data just as an animal's nervous system does. The difference is that the data manipulated, in the case of a human being, includes linguistic data, and an animal's does not. But the consciousness of both person and animal is composed physically of precisely the same elements and mechanisms.

A crucial difference between the kinds of mental operation of humans and animals is that, by virtue of self-consciousness, humans may be aware of a consciously formulated intention behind any orders issued to the body.

This fact in turn seems to imply that a true autonomous being, in some way different from the rest of the organism, actually inhabits the body. Clearly the ancient philosophical problem of free will and determinism is relevant to this supposition but we will have to reserve discussion of it for a later chapter.

The human being's feeling of being able to order the body to perform actions and then to be able to monitor the way that these orders are carried out has made it possible for some philosophers to believe in 'dualism', which is the doctrine that the human mind is composed of a different kind of substance from the body and governs the latter by mysterious or even supernatural means.

Yet how, such theorists have asked, is it possible for an immaterial mind to initiate action in a material body? The truth is that this apparent contradiction only seems to set a problem to philosophers when the mind being considered is a self-conscious, human one. No-one thinks of a cow, or an ant, both of which have minds generated by brains that are somatically identical with our own (though, of course, smaller), as having an immaterial mind inhabiting a purely physical body. No-one suggests that a cow's or an ant's mind is made of a different, more ethereal, substance than its body.

In truth, there is no significant difference as regards either substance or function between a human mind and an animal one. The only true difference between the minds generated by these different life-forms is that the neuro-cerebral system of a waking human being normally generates an entity consisting entirely of organized language and it is this linguistic entity which harbours the notion of 'me' or, extended in space-time, 'the story of me'. But this dependent organism consists purely of verbal data and for this reason is actually no more or no less 'immaterial' than any other data that is processed within a human's or an animal's neuro-cerebral system.

Thus a cow moving towards a more succulent patch of

grass in response to electro-chemical sensory perceptions that there is better pasture in that direction, is behaving in exactly the same way as a young man running to catch up with his girlfriend whom he has just spotted ahead of him. The only objective difference will be that the animal's non-linguistic consciousness will have no knowledge of intention while the young man's linguistic self-consciousness will seem to be initiating autonomous action.

The cow will not be able to understand the reason or the mechanism for its body's change from immobility to motion. Indeed, it will not even be aware of it. The young man, on the other hand, will seem to himself to have, almost magically, issued an order (however perfunctorily) to his own body to perform the physical act of running. But the physics and chemistry in both cases will be identical and no scalpel or electronic or magnetic scanning device would ever be able to expose the self-conscious human being within the purely animal neuro-cerebral system.

Why should it be felt that a person has, or may have, a dualistic nature while a cow or an ant does not? The reason is that the archival memory of the human being, which is richly stocked with language, projects a continual verbal image of the self into the person's conscious mind and this generates the subjective illusion that the person's body harbours what seems to be almost a separate entity or being within it.

Needless to say, a human being is not really composed of an autonomous 'person' occupied with 'thought' while the insentient body patiently awaits its orders to act. In truth, both human being and animal consist mentally of a system of continual stimulus and response governed by sensory monitoring of the perpetually changing environment. However, the intervention of the language circuit in the human being, which provides the illusion that the person is, at least to some extent, autonomous, seems to confer on the person intentionality that is lacking in the animal.

Let us look more deeply into this subjective human sense of'self'. Speaking from my own experience, I can say that while I seem to myself to control my body by volition I do not consider that my relationship to it is similar to, say, my relationship to my car. It is not a case of a living thing manipulating a lifeless one. It is rather a sense of occupying two levels of 'connectedness' with my body.

In the first place I feel that I inhabit it and that I am essentially a virtually autonomous being situated just behind my eyes. But I am also aware that I am an organic participant in all my body's activities and adventures and suffer inescapably from all its woes and malfunctions. My body therefore seems to me, in a political analogy, to be a kind of 'greater me' while the entity behind my eyes serves as the capital where the major administrative departments of state are located.

Most people regularly use phrases like 'my' hands and 'my' stomach and even 'my' brain. But if, as seems to be implied by the possessive case, 'I' see myself as being somehow separate from my hands and my stomach and my brain, then how exactly do I conceive the 'I' that possesses these organs?

Reading and observation both suggest that most normal people will share the author's feeling that while 'I' know that 'my' hands and 'my' abdomen really are parts of 'my' body I do not necessarily think of them as being part of the essential 'me'. Rather, I feel that the core entity, the self-conscious 'me', is a semi-autonomous intelligence which controls, largely by volition, the relatively insentient remainder of my body.

This perception is, of course, precisely what one would expect to find if, as I have maintained, the human mind is really a verbal construction that shares a nervous system with an unselfconscious animal. We thus again receive confirmation that we are 'beings' made of language — language that is invisibly coded into the data-storage facilities of the brain. Such storage of experience is a process common to animals and human beings. But humans possess a crucial faculty which no animal other than the human animal possesses: linguistically-generated self-consciousness. And it is this self-consciousness which gives us the sense (and we must reserve consideration of its validity until a later chapter) of consisting of an executive mind and an executant body.

During the long ages of primate evolution, an animal mind composed exclusively of 'sensons' (particles of sensory data) has gradually become larger and more complex. In phase with this development primate communication has evolved from a set of simple, auditory signals into an elaborate syntactical structure capable of devising and conveying concepts of any kind and of any degree of complexity. In this way, a fully human mind composed of 'mentons' (particles of sensory data fused with, and tagged by, language) has slowly come into being.

As in many other aspects of evolution, and as we implied when discussing infantile amnesia, the governing principle of the relationship of language to mind formation is: ontogeny recapitulates phylogeny. This classic doctrine expresses the fact that the history of each individual may repeat in some branch of its development the history of the species. Thus a baby learns to speak in one to two years. By so doing it recapitulates over a tiny (in evolutionary terms) period of time the development of an animal mind lacking language into a human language-based mind, a process which has taken some three billion years.

Our minds seem to us to be stable and more or less seamless but they are in fact ceaselessly metamorphosing clouds of 'mentons'. These are serially generated, bound into intangible coherence and result from the turbulence produced by the confluence of two data streams. One of these streams is composed of incoming sensory impressions and the other of relevant past events stored in the archival memory. Each human mind, generated by the linguistic component of the data stream issuing from archival memory, thus floats in a medium of meaning-activated particles — mentons — in a way that somewhat resembles a holographic head floating in a confluence of laser beams.

Actually, of course, the image of the self that floats within the cloud of data particles, and which seems to us to be the core-image of ourselves, is not continuous at all. It flickers on and off like the image on a television screen. However, the oscillation is probably not produced in discrete, still images like a film moving through a projector or by continually being redrawn like the image on a television screen. It seems unlikely that either of these techniques of human technology would be sufficiently flexible, or capacious, for processing the vast amounts of data that must be sifted and endowed with coherence ten times per second.

The oscillatory function is more likely to be located in the general cerebral circuitry which 'transports' the data particles to their 'point of confluence' at the present moment. This oscillation is probably produced by the chemical neuro-transmitters acting as transistors and regularly isolating the data cells which comprise the latest evolutionary phase of mind from all other data cells. The mind is thus continually mutating but it also retains a staple identity.

Such a mechanism is highly complex and operates with a speed that seems far more rapid than conscious human thought. But an analogy may help us to accept its possibility. When we look at the image on a TV screen we see a world apparently moving at the same tempo as the world about us. However, we know that this is an illusion. The relatively stately motion of the on-screen action is being generated by elaborate particle movements which are far too rapid to be perceived by the human sensory apparatus.

Largely for this reason, no-one could infer the picture on a television screen from an analysis of particle movements within the set. In the same way, no-one could infer the content of the human mind from an analysis of the molecular-level activity of the brain. Moreover, if the screen of the television set were hidden no-one who did not already understand the technology could infer that the apparatus was designed to generate a picture.

John Searle, the eminent American philosopher, has maintained that it would be impossible to determine the function of a very much simpler device, a chess computer, from an analysis of its electronics. As regards the stupefying complexity of the brain, no-one outside the skull of another person could infer that the neural circuity generates a coherent and self-conscious mind out of linguistic data.

It is for this reason that 'bottom up' (observation and analysis of the detailed structure and behaviour of neural cells and tissue) attempts to determine the relationship of the brain to the mind will never penetrate the sheer complexity of the mechanisms involved. But once a workable theory of the mind, such as the Twin-Data-Stream Theory elaborated in this book, has been tested and proven then it should become possible to trace its constituent parts and circuits.

In the case of both men and animals, mind is best described as a coherent data field. Such an entity can, in this context, be defined as the organized field that results from the turbulent confluence of two data flows meeting at the present moment. It is, however, essential to bear in mind that the turbulence is not itself either mind or consciousness. The initial turbulence must first cohere into a rational state. How can this possibly happen?

It is dauntingly difficult even to imagine such a process. The Twin-Data-Stream Theory maintains that ten times per second a stream of sensory impressions from external reality collides and fuses with a similar but, despite massive screening, vastly greater stream of matching impressions evoked from the archival memory. The second stream bears, in addition to stored sense impressions, a tremendous amount of verbal material selected for its relevance to the sensory imput. Within a few milliseconds, the swirl of data particles that results from this collision has cohered into a meaningful phase of consciousness complete with a rational, or semi-rational, verbal commentary.

This seems to be a sequence of operations of almost stupefying complexity performed at well-nigh unbelievable speeds. Yet while the rate of generation does seem very fast as compared to normal mental activity, it is quite slow as compared to particle-level processes. At the submicroscopic level a thousand millionth of a second can suffice for the entire lifespan of a basic particle. A computer can perform hundreds of millions of calculations in a single second.

It is therefore not too challenging to accept that, even at the much slower rates of cerebral electrical propagation, a cloud of data particles could, in a matter of milliseconds, organize itself into a coherent 'meaning cluster' that included a subjective verbal commentary. This commentary would, in the time available (a tenth of a second), consist of only a word or two, or perhaps a single syllable, but this would be precisely what one might expect in order to give flexibility and fluency to human utterance. Our verbalization must, of course, be geared to a perpetually evolving environmental situation.

Let us try tentatively to imagine this process in a simple, life situation. A man opens his front door and looks out into the street. Instantly all his senses transmit information from the scene before him into his brain. This information is 'pure' in the sense that it is exclusively sensory. He sees (and may in some cases hear or smell or even touch as well) things like cars, buildings, people, sky, air, trees and so on. The data stream from these, and other, sources evokes a complementary stream from archival memory. This second stream is the context-selected and screened residue of all the individual's past experience that is relevant to the present sensory input.

Thus at the moment of the meeting of the two data streams, which is the ongoing present moment, the man's consciousness consists largely of the fusion of the data contained in them with the addition of a substantial stock of verbal material. The archival memory has transmitted towards consciousness words, sentences, remembered remarks, fragments of conversation and texts, possibly summaries of whole books. The vast bulk of this verbalization will, of course, be screened out before it can reach consciousness.

The next stage, which occurs within milliseconds, is for this promiscuous assembly of data to 'cohere' into an organized phase of consciousness. There emerges by means of some electro-chemical ordering process, the nature of which will have to be experimentally determined, a coherent 'molecule' of mind. At the macroscopic level, and necessarily spanning many phases of consciousness, this process might result in our fictional subject musing to himself: 'What a lovely morning. Now where did I leave the car last night? Oh yes, there it is. New green on the trees. And here comes the postman. I'd better take in the milk before . . .'

Our subject is, of course, himself an organism composed almost entirely of language. He governs the animal body by means of linguistic instructions and memories. The somatic tissue of the animal stripped of its language component would be analogous to a chimpanzee or gorilla but the addition of the macroscopically undetectable language transforms it into a reasoning, responsive and creative being.

The time has perhaps come to lay aside that cloak of modesty which so becomes an author and boldly assert that it has always been likely that it would be a professional writer rather than scientist that would ultimately uncover the mechanism of consciousness. This is because a writer's whole working life is a preparation for the task. Since starting to write this book, I have myself perceived that, although I have normally used language simply as a working tool, I have also been intermittently fascinated and puzzled by its almost limitless potential and by its relationship to the human mind and to objective reality.

If indeed the Twin-Data-Stream Theory proves to be right, and the linguistic nature of human self-consciousness becomes generally accepted, it will be recognized that authors, including bards working in an oral tradition, have probably been wrestling with the problem of consciousness since before writing was invented. The greatest of all writers, William Shakespeare, epitomized the conclusions reached in the present book in the splendid phrase 'thought's the slave of life'.

In the last century or so, naturalistic literature has become more and more adept at reflecting the essential nature of consciousness and in modern times it has achieved superb representations of that 'story of me' which, as we have seen, is essentially the human mind. This life saga, which each one of us tells him or herself from early childhood until death or some other physical impediment truncates it, is a ragged, heterogeneous, often non-sequential yarn very different from the polished stories told by most professional novelists.

The 'story of me' jumps about in time and tense. It is sometimes repetitive or non-sequential. It includes snatches of dialogue, song, self-exhortation, bar-room jokes, exclamations of surprise, delight or agony, orders to children, denunciations of perceived enemies and indeed every kind of spoken or written utterance that human beings have ever produced. It is peppered with exclamations and irrelevancies. But for all its crudeness and shapelessness it has drive, pertinence and comprehensiveness far beyond the scope of any invented narrative.

And yet, over the last hundred years or so, a few profound and deep-probing authors have succeeded in producing impressive fictional versions of this immense yarn. In order to do this they have spontaneously invented a literary form which has become known to criticism as the 'interior monologue'. This is a technique which was probably first used by Leo Tolstoy (although Shakespeare's soliloquies foreshadow it) in order to give a sense of both the chaos and the organization that characterize the mind of the eponymous heroine of his great novel *Anna Karenina* as she prepares to commit suicide. We have seen of course that chaos, followed within milliseconds by coherence, represents the normal ten-times-per-second alternation of human consciousness.

Probably the most extensive and compelling example of an 'interior monologue' that has ever been penned was produced by James Joyce. It forms the final section of his immense and often apparently chaotic (although actually profoundly ordered) work *Ulysses.* In this novel, Joyce explores the consciousness of another heroine, Molly Bloom. Possibly both Tolstoy and Joyce intuitively felt that the female, as the custodian of human continuity, was the appropriate sex to provide the first literary expression of what I have called 'the story of me'.

The fact is that the closest we can come to defining the essence of consciousness is to state that it is what Tolstoy and Joyce intuitively perceived it to be. They imaginatively recreated the subjective stream of mentation in order to endow their characters with heightened and intensified reality. By so doing they generated narratives that mimick the 'stories of me' which virtually all verbally-equipped human beings produce and which constitute an apparently autonomous power within the body of the human animal.

The truth is that the human mind is a verbal entity, a serially generated linguistic commentary concerned with every aspect of the organism's existence. It defines present experience, compares it to past events and thus suggests the best course of future action. To the (vast) extent that we are different from all the other life-forms that inhabit our planet it is because we possess verbal self-consciousness. This achievement enables us to keep track of ourselves and of our adventures in time and in space and to modify our own evolution and ultimately our destiny in the universe.

Human self-consciousness is the only kind we have thus far encountered. Human beings are the only kind of 'beings' of which we have any knowledge. While horses and pigs exist, and dinosaurs once existed, none of these, or any other life-forms of which we have any knowledge, can know that they exist or be aware of either space or time. It was only with the advent of language, and hence self-consciousness, that Chronos was born. Before the birth of Chronos, of time, the unnamed ages flowed by unrecognized.

My simple, linguistic experiment with the flock of sheep indicates that it is possible for us conceptually to seat fellow inhabitants of our planet more firmly within the framework of evolution just by giving them a more resonant name. But the creatures themselves remain stubbornly anonymous. It may be true that a dog which we call Rover will soon learn (by conditioned reflex) to associate the twin-syllables of the word Rover with the authority of a summons but it can never know that it has a name or that the sound it hears is a name or the purpose of names or indeed anything whatsoever about the cosmos that it inhabits. This is because all 'knowledge' is crucially dependent on language. Without language it is quite simply impossible.

Over long spans of evolutionary time, and if we use our *de facto* suzerainty of the planet wisely, there may come into existence — although the possibility seems very remote — true language-equipped ovine beings or bovine beings or porcine beings. At present it seems that the most likely candidates for neighbours with whom we could converse would be cetaceous beings. But for the foreseeable future the only names the huge labyrinth of non-human creation will bear will be those we bestow upon its numerous forms. For this very reason, we have a moral and ethical imperative to assume the role of representatives and guardians of life on earth since no other species, lacking language, can look after its own interests.

This is clearly an awesome responsibility and how successfully we perform it may ultimately demonstrate whether we are worthy to bear the title (on a level slightly below that of deity but well above that of sheep) which we have graciously bestowed on ourselves, that of 'human(e) beings'. It is conceivable that we will one day make contact with other self-conscious intelligences elsewhere in the cosmos (although our attempts to date to do this have proved dismal failures). But it seems most likely that we will have to wait until a species on our own planet achieves syntactical verbal language and starts to tell itself the 'story of me' before we are able to share the experience of being a 'being' with another being.

We embarked on the voyage of discovery charted in this book with a dream and we have ended it with a story. I hope that I have convincingly demonstrated in these pages that our minds really are narratives generated by the brain. It is a challenging concept (some people may indeed find it rather a disturbing one) but the truth is that to the extent that we are more than, and different from, all the other creatures on earth we human beings resemble narrative — written stories in books or tales told by a story-teller — more than we do a biological entity. In *our* beginning, certainly, as well as in our continuance, is the Word.

# PART TWO

# Malfunctions

What, if anything, is the relevance of the Twin-Data-Stream Theory to pathological mental states? Clearly anything approaching an exhaustive answer to this question is far beyond the scope of an introductory account such as this. It is important, however, to provide readers with at least a glimpse of the relationship between the theory and some of the more prevalent disorders of the human brain and mind. The chief purpose of this chapter, therefore, will be to discover if the theory has anything of value to contribute towards understanding, and thus potentially alleviating, a few well-known mental disorders.

Since dreams were the key to the discovery of the Twin-Data-Stream Theory, let us start by looking at a relatively mild malfunction of the sleeping process itself. This is sleepwalking. It is far more common than many people suppose. A modern Swedish survey has found that the high percentage of 75 out of 212 children — more than a third — experience at least one episode of sleepwalking.

Empson describes a typical example of this somewhat eerie phenomenon as follows: 'after getting out of bed the sleepwalker may get dressed, and then walks about, often repetitively, or may remain standing still. He or she typically returns to bed spontaneously after as much as 30 minutes of activity, and remembers nothing about the incident . . . Sleepwalkers will respond to verbal suggestions, but do not engage in coherent conversation.'

The Twin-Data-Stream Theory accounts for this strange phenomenon by stating that it probably involves the same mechanism as that which causes dogs to move their legs in a running motion during REM sleep. In other words, human sleepwalking represents a mildly pathological involvement of motor circuitry with the downloading of visual sense impressions into archival memory. A dog that is engaged in filing into its archival memory stored memories of running may move its legs appropriately in sleep. A sleepwalker filing stored memories of walking does the same. Empson observes that a sleepwalker resembles 'a malfunctioning automaton'.

Now this intuitive insight actually represents a powerful confirmation of the Twin-Data-Stream Theory. This is because the theory maintains that a sleeping person is, in fact, a kind of 'automaton' in which normal, diurnal control mechanisms, linked to sensory scanning of the environment, have been 'switched off'. A sleeping person walking about without a functioning mind therefore can with considerable validity be thought of as a 'malfunctioning automaton'.

The mythic image of sleepwalkers in the popular imagination often includes scenes in which they perpetrate great, and perhaps fatal, violence. This suggests that sleepwalkers are commonly perceived as being in some sense 'inhuman'. In actual fact I have been unable to find any reference to actual harm having been committed by a sleepwalker. This finding corresponds to the reality postulated by the Twin-Data-Stream Theory which regards sleepwalking as a precarious and largely uncontrolled involvement of motor circuitry. Such undirected action could only fortuitously do any harm. We thus see that both the sinister popular image of a sleepwalker and the harmless reality tend to confirm the Twin-Data-Stream-Theory's view of the condition.

Schizophrenia, with its strange temporal distortions, its voices, its delusions and hallucinations, its personality disorders and perhaps most conspicuously its degradation of the afflicted individual's orientation in the world, is a pathological state peculiarly suitable for examination in terms of the Twin-Data-Stream Theory and especially that aspect of it concerning screens.

These screens remain to some extent hypothetical. The author has, nevertheless, had various experiences that strongly confirm their physiological existence. To give a typical example, while washing a wooden salad bowl quite recently, I was struck by the object's harmonious form. Perhaps half a minute later there came into my mind an image of a wooden fruit or salad bowl seen, almost certainly once only, at the age of perhaps five or six.

This demonstrated to me, not for the first time, that the data retrieval mechanism of the mind is capable of returning to consciousness a many-decades-old image from the millions stored in archival memory and introducing it into consciousness very rapidly. Rapidly, yes, but, as compared to the virtually instantaneous appearance of many such evoked memories, a delay of half a minute testifies to a huge amount of data sifting.

I tried to discover what there was about the remembered wooden bowl which had caused its image to be 'despatched' from archive into consciousness at that moment. Having, like most people, handled wooden bowls from time to time throughout my life, I was aware that my archive must hold a vast store of memories of them. Why had that one memory alone been 'reanimated'? Then I vaguely recalled that at the time of the original childhood experience, someone had commented on how attractive wooden bowls are. This observation might have made a strong impression on a very young child striving to find reasons and context for a dawning aesthetic apprehension.

It is possible, therefore, that when I started washing the wooden bowl thousands or perhaps millions of items from archival memory, concerning innumerable experiences of wooden bowls, and with ramifying cross-indexing embracing subjects varying from carpentry to archeology, would have been 'alerted' for possible despatch into consciousness. But conceivably from amongst all these references only a single one concerned the aesthetics of wooden bowls and it had been the aesthetic qualities of the bowl I was washing that had especially struck me.

As regards the function of the screens in schizophrenia, let us first consider the voices and other hallucinations which are one of the key features of the disorder. It is known that something analogous to them can result from extreme sleep deprivation. In the case of protracted insomnia such 'hallucinoid' phenomena are known as 'waking dreams'. But in fact waking dreams are, as we shall see later, different from true schizophrenic 'hallucinations'. Waking dreams derive from a very rare condition but one in which screen malfunction is not essentially implicated.

An individual suffering from waking dreams is in a state of desperate sleep deprivation. As a result, he or she ultimately starts to dream while remaining at least partially awake. While very rare, the disorder classified as 'waking dreams' is recognized by sleep researchers. The reason for it, however, is not. In terms of the Twin-Data-Stream Theory the bizarre process would be caused by too much sensory data reaching the short-term memory as a result of lack of sleep and finally exceeding the available storage capacity. Ultimately the overcharged short-term memory would simply 'overflow' and initiate in the waking mind the cross-indexing and filing process into the archival memory which, when the brain is functioning normally, occurs only in sleep.

Meanwhile some degree of consciousness would continue to be generated by the incoming data stream of sense impressions. Perhaps too a certain amount of evoked data from archival memory would surface despite the competition from the filing process that was now taking place. The result of this unnatural combination of nocturnal and diurnal activity would be 'waking dreams'.

Screen malfunction may ultimately play a part in the production of waking dreams but it is not a primary cause of them. Such malfunction is central to schizophrenia. The schizophrenic mind, like all minds, weaves through space and time generating a continuous input stream of sense impressions. These impressions transmit to archival memory a demand (actually they themselves constitute that demand) for elucidation and guidance. Archival memory selects from its vast store of referential material all that is relevant and despatches it up through the screen cascade.

But in the case of schizophrenia, the screens fail to do their job properly. They are to a greater or lesser extent disabled and allow too much data through into consciousness. The schizophrenic mind therefore receives vastly more data than is useful, or even usable, for providing guidance — so much, indeed, that it cannot even be rationally integrated into consciousness. Purposeful action in obedience to the life situation perceived by sensory scanning, which is what the whole system aims at, is swamped in a cataract of data. It is this unmanageable torrent of surplus data that manifests itself to schizophrenics, and psychiatrists, as hallucination.

Let us suppose that when I was washing my wooden bowl, instead of a single chaste memory of such a bowl coming into my mind from a childhood experience, my consciousness had been besieged by a swarm of references. Suppose I had received fleeting and shifting visual impressions of long production lines of wooden bowls, armies of carpenters sawing and turning in immense workshops, forests being hewn down by huge machines, arrays of wooden bowls in shops, kitchens and living rooms etc. Would not such an intimidating visual assault have proved a distressing experience? Could it not reach an intensity that would cause both sufferer and any attendant therapists to conclude that the victim was 'insane'?

Now let us consider the auditory hallucinations of schnizophrenia. These are perhaps the defining characteristic of the pathology. The schizophrenic very often hears voices and these are primarily hostile and condemnatory. This inherently nightmarish situation would come about because the incoming stream of sense impressions would introduce into consciousness an aspect of external reality which required verbal response or guidance. Such context-relevant data would then be automatically summoned from archival memory but instead of highly selective and pertinent verbal material (literary and anecdotal) being selected out by the screens for final despatch into consciousness screen failure would mean that the mind was besieged by harsh and disagreeable 'voices'.

Some of these voices might indeed be cruelly hostile and critical. This would paradoxically result from the fact that, for a human being, constructive verbal criticism is probably one of the most valuable functions of archival memory output. But in the case of schizophrenics, the screen failure would mean that instead of a few helpful suggestions, or even admonitory remarks, reaching consciousness, the mind would instead be assailed by a torrent of harsh, insensitive and probably only marginally pertinent voices. The unhappy victim would be bombarded by hostile denunciation.

To generalize on what has been said, it is a characteristic of schizophrenics that they tend to avoid importunate stimuli. They do not seek out exciting places or people. They are not addicted to social or group occasions. On the contrary, schizophrenics tend to crave quiet and calm. Perhaps the definitive statement of the ordeal of the schizophrenic mind has been made by John Clare, the nineteenth-century poet who died in an asylum and who is now considered a classic early example of the modern schizophrenic syndrome.

Clare's notion of ultimate happiness, as expressed in his great, despairing poem known simply as 'Written in

Northampton County Asylum', was to lie alone on the grass beneath the sky, in other words to be situated in as near a stimulus-free environment as could easily be found. Perhaps significantly, part of the yearning expressed by the poem is to sleep 'full of high thoughts, unborn' which may be interpreted as free from the torment of unwanted, unsummoned and chimaeirical imagery.

There is a special form of schizophrenia which is known as cataleptic schizophrenia. In this state, the subject freezes physically, and as far as possible mentally, for long periods of time. He or she may maintain a posture — become a living statue — for hours or days on end. In terms of the Twin-Data-Stream Theory the cataleptic represents a sufferer who has responded to the torment of the excessive stimulus resulting from screen malfunction by retreating into what he or she conceives as the last bastion of sanity: complete immobility.

He or she attempts to get rid of the tormenting, unscreened, or inadaquately screened, data stream by banishing all movement of mind and body, perceiving intuitively that if he or she can only shut off, or at least weaken, the sensory input stream this alone will reduce the summoned-up archival memory stream.

It is a doomed attempt because the cataleptic, despite his or her rigid posture, is still the subject of sensory data stream input which will be almost as intensive as in normal circumstances. It has, in fact, often been noted that cataleptics, beneath their eerie immobility, remain fully aware of all that goes on in their vicinity. Occasionally they are slapped or otherwise maltreated by conscienceless attendants and at some later time they may reproach the perpetrator. The patient is, however, prepared to tolerate even physical assault rather than abandon immobility and risk increasing the terrible onslaught of the unscreened, or inadequately screened, data stream.

In an earlier chapter I contrasted schizophrenia with Korsakov's syndrome. I suggested that these two conditions are opposites. The schnizophrenic receives too much data, as a result of screen malfunction, while the Korsakov's victim receives too little and is driven desperately to improvise the ongoing 'story of me'. The two states, however, are not really exact opposites since the Korsakov victim does have a functioning archival memory although it is a memory that was stocked years or decades in the past.

The true opposite of schizophrenia, in terms of screen malfunction, is in fact autism. Strangely enough, when the austistic syndrome was first observed and defined just after the Second World War it was considered to be a kind of infantile version of schizophrenia. It is actually a distinct syndrome with very different symptoms. The Twin-Data-Stream Theory attributes this error to the fact that schizophrenia and autism are not adult and juvenile versions of the same disorder but, as regards etiology, mirror images of each other and therefore each implies the other. By now autisim has been generally recognized as a distinct mental malfunction.

Autism is characterized by a large range of symptoms which can vary from dreadful states such as almost permanent convulsions and seizures to the milder, but still desperate to sufferers, complete withdrawal and lack of emotional expression, inability to speak and so on. But the key defining symptom, which was apparently intuitively perceived by the psychiatrists who first named the condition, is that autistic patients seem to have an eerie remoteness and detachment. They give the impression of being walled into themselves and to have hardly any interface with events or indeed with any aspect of the world of normal experience.

Autistic patients, like schizophrenics, suffer from screen malfunction. However, the malfunction is the direct opposite of that afflicting schizophrenics. Instead of too much data pouring through the screens into consciousness there is too little. The degree of severity and specific manifestations of autism probably depend to some extent on how large or small a quantity, and perhaps what type, of impressions actually do get passed by the screens.

In the very worst cases of autism, however, it seems likely that very little or no data of any kind gets through the malfunctioning screens. Such an autistic sufferer would thus only have, in effect, half a mind. Data would pour into consciousness from the outside world but would fail to evoke any complementary guiding, interpreting and unifying stream from archival memory. There might, in fact, be no, or only a rudimentary, archival memory present.

The mind of a person afflicted in this way must be ineffably strange as well as tragic. It would consist of an endless invasion of incomprehensible impressions. Deliberate thought for such a sufferer would be like striving to put together a surreal jigsaw puzzle, the pieces of which would indeed show fragments of visual reality but which could neither be physically joined up nor arranged in a way that suggested any large-scale picture.

Extreme autistic cases are those unhappy sufferers who spend their childhood rocking desperately in the attempt to obtain some kind of self-awareness separate from the blizzard of meaningless (because they are never interpreted by feedback from the archival memory) sense impressions. As they grow older and if there is no amelioration of their condition a proportion of such victims will end up as voiceless 'idiots'. Such unfortunates never manage to generate a functioning mind.

But if, as happens in some cases, the screen defect mends itself and a flow of data from archival memory starts up, then these same near 'idiots' can suddenly blossom into thinking, feeling people. Even in such relatively fortunate cases, however, the long period during which mind formation has been non-existent or grossly disordered makes it unlikely that the autistic sufferer will ever be fully normal.

Let us turn from what are clearly recognized as pathological states to another which, while posing severe problems in coping with the everyday world for those who are afflicted by it, is often thought of as a kind of superhuman condition. This is the state of those apparent wonder-workers known as 'idiots-savants'.

This group of people has great powers but also great disabilities. The very name expresses their curious centaur-like nature. It is a French phrase but works the same in English. It refers to human beings who are 'idiots' in the arts of life but who are also 'savants', which the Oxford dictionary defines as people 'of learning or science'. The term 'idiot-savant' is a disagreeable one and is being abandoned in favour of more sympathetic phrases such as 'calculating wonder'. However, since the original term is so apt at representing the condition, I propose, if reluctantly, to continue to use it in the present section.

Can the same person be an 'idiot' and a man 'of learning or science'? The answer is: no, he cannot. In terms of the dynamics of the human mind, the combination is impossible. An extension of this immediate truth is that while 'idiots-savants' really are, in a sense, 'idiots' they are never genuinely 'savants'. They do not make — are incapable of making — important contributions to our understanding of the world. Rather they perform mental tricks which are analogous to physical ones such as juggling or tightrope walking.

These mental feats may, nonetheless, be awe-inspiring. The characteristic achievements of idiots-savants are usually in the field of mathematics. To give just one example, the neurologist and author, Oliver Sacks, documents a pair of idiot-savant identical twins one of whose favourite pastimes was to sit swapping six-figure prime numbers. These immense numbers, only divisible by one and by themselves, take computers a long time to locate. The twins could find them in moments.

Other idiot-savants have incredible musical, or even artistic (considered as the ability to pen highly evocative drawings from an immense visual memory store) powers and they may be endowed with other impressive abilities. But they all share the same negative quality. They are unable to look after themselves in the real world. Sacks' calculating wonders, for example, were observed giggling and swapping their prodigious primes in the little room they shared in a mental institution. The question naturally arises: is there some connection between the superb powers of 'idiot-savants' in their special fields and their helplessness before the everyday, and apparently trivial, challenges of life?

The pathology of idiots-savants almost certainly derives from a fault, or an interlocking sequence of faults, occurring in one or more of the stages involved in data processing: sensory-input, initial short-term memory store, downloading into archive, storage and modification in archive and finally despatch and screening of the data stream from the archival memory into consciousness. In essence the problem is that the data processing system regularly, and without conscious involvment, selects for one minor, or relatively minor, aspect of the incoming sensory data stream.

The majority of idiot-savants — i.e. the calculating wonders — would thus preferentially select for mathematical data. There is abundant evidence that the peculiar mental orientation of idiot - By the time a particular individual has reached physical maturity he is endowed with a vast memory archive of mathematical information but has little useful data of any other kind. As a result, the over-specialized individual, who can effortlessly pluck six-figure prime numbers from his data store, so far from being a kind of mental superman is actually a grievously limited person. He may even be unable to perform tasks like tying his own shoelace because he does not have enough back-up data stored in his archival memory to guide him in performing the job.

What the idiot-savant really represents, therefore, is not a person with an incredibly capacious mind but a deprived being with a very limited one. Tying a shoelace, ordering a cup of coffee, writing a note to the boss, doing the shopping and even picking up a fallen coin or pin are all immensely demanding tasks that dwarf the relatively limited, because highly specific, one of finding big prime numbers.

To tie a tie or a shoelace properly depends on being able to summon up cataracts of data from archival memory that will inform every muscular coordination and sensory investigation that the process entails. In order to undertake successfully what strikes 'normal' people as routine and everyday tasks requires vastly greater information stocks, and information processing capacity, than the deprived idiot-savants require for performing what seem to be their far more impressive feats.

A useful analogy here is with a computer. Programmes have now been devised that enable computers to play chess at world-championship level. On the other hand, no-one has yet devised a programme that makes it possible for a computer to set out a box of jumbled chess pieces onto the board with as much skill as, say, a four-year old child. At first consideration it seems obvious that playing high-level chess is a more impressive achievement than putting a few pieces of wood into their proper places on a tray. But it is not so. The quantity of data, and especially of coordinated heterogeneous data, required for the latter task vastly exceeds what is needed for playing the game extremely well.

Sequenced into digital instructions it is almost impossible to provide a programme that will enable a machine (equipped with the proper mechanical manipulators) to take a collection of jumbled chess pieces, colour-coded into two different sets, and transfer them in an orderly manner to their correct places on a chequered board. The task involves detailed sequencing of three-dimensional movements, discrimination of shape and alignment of shape and colour with destination as well as nice judgement of where precisely and with what degree of force to grip each piece. All this complex and interlocking information can be held effortlessly in the mind of a four-year-old child.

It is perhaps also instructive, in this context, to compare the difference between animal and human performance of simple tasks. Take a dog looking for a thrown ball and a person doing the same. The dog will rove through the rough grass to approximately where it last saw the ball, paw about a bit, stand irresolutely, start to trot back to its master, be urged to return, run about aimlessly and possibly ultimately find the ball by chance or never find it at all through becoming distracted by some irrelevant factor.

A person will, of course, make a systematic search and take into consideration background guidance data, fed to him or her from archival memory, bearing on factors such as the direction of travel of the ball, its velocity, the nature of the terrain and so on. The human being may not find the ball any more than the dog does but the crucial consideration is that watching the two different life-forms at work clearly reveals that vastly different mental dynamics are being brought to bear on the task.

The dog is unable to derive from its rudimentary archival memory much if any guidance as to search techniques and hardly even a consistent memory of the ball itself (it may ultimately pick up an old shoe and take this hopefully to its owner). The man is being informed by a lifetime of experience of the huge potential array of responses that can be made to even such a simple challenge. He is drawing in his search on the language-tagged guidance data accumulated throughout his life although he is not aware that he is doing so. When an animal, such as a sheep-dog, behaves with apparent purpose it is really responding to sensory cues (whistles in the case of the sheep-dog) from a human being which break the overall task down into simple components that can be held in the animal's non-verbal memory archive.

So we see that, looked at properly, it is the ordinary

person going about his ordinary business and whose information processing system is geared to coping with experiential reality, who is the true miracle-worker rather than the 'calculating wonder'. For the ordinary person has available, and exploits at every fleeting physical and mental moment, an immense library of cross-referenced data. The idiot-savant, on the other hand, is the sad and underprivileged possessor of a mind that has no access to the riches held in archival memory which 'ordinary' people routinely deploy.

The aim of this chapter is not to demonstrate that the Twin-Data-Stream Theory can analyze and help to solve all psychopathological problems. Clearly it cannot. Moreoever, there are many pathological conditions, such as hysteria, manic-depressive disorders, migraine and others, for which the theory has, in the present state of understanding, little guidance to offer although it is not incompatible with them. But it is worth briefly analyzing the theory's relevance to a condition that is not pathological but is curious and has apparently baffled many commentators.

This condition has, moreover, attracted a great deal of attention in recent years from believers in the possibility of paranormal and even supernatural realms. It is known as 'out-of-body' experience. We will concentrate in this short section on just one often-reported example of it. In this type of experience, a patient undergoing a surgical operation finds that while the operation is taking place he has floated up from his body and, from a point of view near the ceiling, has found himself able to look down on both his body and the medical team at work on it.

Many readers will doubtless realize at once the explanation that the Twin-Data-Stream Theory provides for such an alleged phenomenon. It is that the patient, in anticipation of surgery and possibly informed by some knowledge of surgical procedures, has brooded a good deal about his approaching ordeal. By so doing he has stocked his short-term memory with imagined images of the forthcoming operation. Then when he has been anaesthetized for the surgical procedure, he would have begun to download impressions from the preceding waking period. During the brief time of awakening, either while still on the operating table or later in the recovery room, he simply dreamed the 'out of body' experience.

The reason why I find this situation of special interest is, of course, because it concerns precisely the same type of sleep experience as the one that originally set me on course for writing this book. This is a dream that involves the dreamer in becoming his own observer. It is, in fact, a 'two-I' dream.

We will conclude this chapter with a look at how the Twin-Data-Stream Theory relates to various forms of amnesia and thereby at the theory's broader social implications. Concerning memory, it is important initially to state that the qualities people conventionally regard as the indicators of the uniqueness of each human being are often unimportant generalities. They frequently have little or no connection with the complex, true individuality which, as a result of having a necessarily unique archival memory, each one of us possesses.

There is a well-known case (about which an excellent TV documentary has been made) of Korsakov amnesia suffered by a BBC music producer and musician which tends to support this fact. The victim's pathological condition resulted from an extremely severe attack of viral flu. After the illness, he remembered virtually nothing of his former life although he did retain his musical ability.

He was, in the characteristic way of Korsakov sufferers, unable to hold onto new sense impressions for more than a few minutes. Nevertheless, he remembered the wife he adored although he would invariably greet her, when first setting eyes on her after even a very brief absence, as if they had not met for an age. It was heart-rending to see him fall into her arms — perhaps when she had merely been in the adjoining room for a minute or two — with the anguished exclamation: 'Oh — darling! I've missed you so much!'

In the television documentary that was made about this unhappy man, he was seen endlessly playing patience alone in his nursing-home room. This was how he passed most of his days, uneasily aware that something was gravely wrong but persuaded, whenever he met new people, that he was just at the point of recovering normal life after a nightmarish illness. He would explain: 'it was just like being dead'. Amongst old friends, who seemed to him like agreeable strangers, he might enthuse: 'This is the first time I've felt well for days — no, weeks'. But, inevitably, a few minutes later he would have forgotten what he had said and have not the faintest recollection of those he had been speaking to.

But the crucial point to be made is that this unfortunate man was, his wife maintained, in essence exactly the same person that she had always known. She insisted that his fundamental personality was quite unchanged despite the cerebral catastrophe that had robbed him of the chance of ever again living normally or happily. The same thing applies even more forcefully to cases of traumatic or psychological amnesia. The victim normally forgets all the particular details which distinguish his life from that of anyone else but remains manifestly the same person.

What exactly is it that such victims have lost? We do not, in fact, have to probe very deeply before realizing that it is basically only the conventional insignia of individuality: name, address, occupation — and so on. If we consider the matter with some attention, we also soon perceive how very marginal such largely bureaucratic information is to all that is essential for defining each human being's individuality. The amnesiac forgets his or her name and address, the names of a handful of relatives and other persons close to him or her, a few formal details and a bit more besides. But this represents only a tiny fraction of the data stored in his or her archival memory. Most if not all of the scanty information that has been lost could be contained on an identity card and all of it on a digital smart card.

But of course we do not need, for our own everyday purposes, to carry digital smart cards around with us because we have in our brains a special mechanism for continually reminding us of the basic data pattern of our particular social origins and orientation. This 'identity check' facility operates on an automatic, or drone, basis analogous to the 'urgent memo' repeat facility ('phone the doctor, phone the doctor') mentioned earlier. Every second, or more, or less, throughout waking life this basic 'identity check' is projected from archive upwards into consciousness. In this way we all go through the world secure in the knowledge of who we really are and without recurring panic attacks about our identity.

The practical, if also metaphysical, question 'who am I?' doubtless occurs to most people from time to time. In such cases, at least the mundane (if never yet the metaphysical) answer usually comes through to them clearly and comprehensively enough. 'Who am I? Why I know perfectly well who I am. I am actually Bill Brown, an insurance salesman and I live at Number 1, History Mansions. I am 41 years old, and I have a wife and three children. I was born in Glasgow . . .' and so on.

Now, probably 99.999% of the times that such an 'identity check' is despatched towards consciousness the screens will filter it out as 'surplus to requirements'. But there will come an occasion, perhaps when a smiling clerk seated on the other side of the desk from Bill Brown asks him politely: 'Could you give me your name and address, please, sir?', when it is useful to have the reply to hand and the metronomic identity check will then prove its value. In such circumstances, Bill Brown will have no difficulty in returning the interlocutor's smile and stating confidently: 'Yes, my name is William Brown and my address is Number 1, History Mansions . . .' and so on.

It is likely that the data cells in charge of the 'identity check' have, because of the heavy-duty nature of the data stored in them, back-ups and that the back-ups themselves have back-ups. This cumbersome apparatus, designed to safeguard the precious data that is perceived as necessary to safeguard our individuality, means that a change in the data resulting from, say, moving house or getting married, often disorientates the individual. He or she may go on dispensing obsolete data for some time before the new, amended 'identity check' is run in and working properly. 'Bill Brown — yes — and I live at Number 1, History Mansions — no, hang on, we've moved — yes, that's right, I now live at Number 10, Relativity Close.'

It may, however, come to pass one unlucky day that Bill Brown receives a blow on the head which knocks out the 'identity check' data. Alternatively, he may undergo such a fearful psychological trauma that it triggers a protective mechanism that shuts down the offending circuitry. In either case, the mishap obliterates, permanently or temporarily, the data that is normally automatically provided by his 'identity check'.

Soon afterwards he may be led into the casualty ward of a hospital having no idea who he is and unable to remember anything about his life before the traumatic event. Doctors may discover, nonetheless, that he knows his way about the city in which he has been found. He may also know the latest world and national news and will, in fact, although the doctors may not be in a position to verify this for some time, be exactly the same person that he was before the trauma except for having forgotten the basic bureaucratic details of his life that are contained in the 'identity check'. His personality, range of knowledge and skills, temperament, tastes, abilities, use of language — all will proclaim him the same man.

If old friends or close relatives are located they will note, whether Bill ever recovers his 'identity check' data or not, that Bill Brown before the trauma and Bill Brown after the trauma are — in all but the trivial sense of having misplaced a few official details — the same person. The crucial fact here is that the 'identity check' data constitutes only a minute, almost negligible, part of the real personality and individuality of the man. Bill Brown's true individuality resides not in the 'identity check' data but in the massive uniqueness of his archival memory. This constitutes an authentic individuality which, as compared to the few, simple facts of the 'identity check' (facts which are merely formulated and then regularly rehearsed) is rooted in the slowly garnered and necessarily person-specific data stocks of a lifetime which individualize each one of us.

Understanding of this fact is relevant to answering a question, deriving from the awesome power and rate of development of our technology, that is quite often posed by philosophers and 'ordinary' people alike these days. This is the question of whether we could one day be superseded by thinking machines as the most intelligent 'beings' on our planet. The reassuring answer, suggested by the Twin-Data-Stream Theory, is that we could not. The human mind, as we have seen, is not merely an inert heap of pre-classified knowledge that might be simulated by feeding codified data into digital memory banks. It is rather a unique, dynamic, individually-harvested, situation-responsive constellation of language-bound data. Each unique archival memory is produced by each unique human being exclusively for his or her use.

Our archival memories are crucially dependent on the minutiae of the actual moment-by-moment experience garnered by the individual as he or she moves purposefully (although with serial purposes) through space and time. It has, moreover, been historically generated by a process that has already (genetically) absorbed and assimilated into the genes of each individual the lessons learned and the achievements made at all the earlier stages of biological evolution on this planet.

We can thus see that simply assembling a mobile robot

and stuffing its 'head' with the world's entire stock of codified knowledge could not produce a 'time mind' analogous to our own. The manufactured artefact could never be other than a more or less effective and more or less adequately data-stocked automaton. Even if we could endow the robot — android, cyborg, mobile computer — with motivations that were akin to instincts, tropisms that were akin to emotions, aspirations that were akin to ambitions and passions that were akin to friendship, love and sexual desire, the end result could never be much more than a travesty of a human being.

It is conceptually impossible for the kind of biological beings that we human beings are even to initiate the evolution of mentally superior beings since our own evolution would always keep us ahead. There may exist, in some other part of the cosmos, a race of more highly-evolved living organisms and these aliens might be capable of building us. It is even remotely possible that, long ages ago, they did. But we could never build them. We are thus secure from the competition of machines in the role of chief custodian of mind on the planet earth. For all that, it would probably be a good idea for us to maintain a fraternal, and yet wary, watching brief on the dolphins.

# Towards a Philosophical Context

Look up 'sleep' in any dictionary of quotations and you will probably find similar ideas about it expressed by the world's greatest poets and thinkers. According to Shakespeare, sleep 'knits up the ravelled sleeve of care'. In another passage Shakespeare calls it 'nature's soft nurse'. Sir Philip Sydney describes it as 'the certain knot of peace . . . the balm of woe'.

Three hundred years later, John Masefield refers approvingly to 'quiet sleep and a sweet dream'. True, several poets, like Tennyson in the phrase 'death's twin brother', make a comparison, which is both obvious and ominous, between sleep and death. Nonetheless the voices dissenting from the notion that sleep is a wonderful thing are few indeed.

Literary sources also suggest that we humans seem to ourselves to be most appealing in sleep. While much of our literature chronicles the turbulence and terror that may afflict our waking lives, in sleep we seem united in simplicity and touching vulnerability. Such considerations apply with special force to the notion of the 'innocent sleeping child'.

But the Twin-Data-Stream Theory substitutes for such endearing images of sleep the concept of an electro-chemical automaton relentlessly engaged in high-powered data processing. I have sometimes wondered if even the Darwinian revelation that we are first cousins to apes can have seemed as distressing to the populace a century and a half ago as the new idea might strike some people today. In this chapter, therefore, I will make a very tentative, brief and probably premature attempt to place the Twin-Data-Stream Theory in a social and philosophical context.

It might be asked: if such a 'soulless' reality as relentless data processing is genuinely the truth about sleep then does it not follow (at least as an implication) that we should seriously consider various other rather dismal theories, often originally derived from science fiction, about ourselves and our origins. One such theory, accorded serious attention by a number of reputable scientists, states that we may not be evolved biological beings at all. Our remote progenitors, according to this idea, would not even have been terrestrial life-forms but something along the lines of industrial robots manufactured by a race of technically advanced aliens making a stopover on our planet or perhaps using it as a vast laboratory for a grandiose evolutionary experiment.

But we need not limit ourselves even to this ambitious level of speculation. On the authority of at least a few philosophers and scientists, we are justified in asking ourselves if human beings really are three-dimensional beings. Do we, in fact, actually experience a physical existence on the surface of a physical planet or could we conceivably be no more than computer-generated simulations inhabiting a virtual-reality environment?

Leaving aside for the present such bizarre notions, I have nevertheless felt some anxiety, while at work on this book, as to how readers will respond to the Twin-Data-Stream Theory's basic idea. This maintains that a 'person', even though fully material, and indeed biological, is not a stable and continuous entity but rather an intermittent and discontinuous one. The theory, as we have seen, postulates that each one of us undergoes a tiny death and resurrection perhaps every tenth of a second. Harder still for some people to accept might be its insistence that, mentally, we all cease to exist at all during sleep.

Can any comfort be offered to readers who might be distressed by such unorthodox views about the nature of the human mind? I feel that it can. In the most general sense, it can be asserted that any revised image of ourselves that the Twin-Data-Stream Theory may seem to call for will still be compatible with traditional notions of what it means to be human. We may lament the loss of, and even develop an enduring nostalgia for, older self-images based on classic ideas of dignity and autonomy but the new theory remains compatible with them.

Our sensibilities, emotions, instincts and intellectual potential, our capacity for generating culture and civilization, our power to love, laugh and suffer remain just as true and as characteristic as they always have been. Bearing this reassuring notion in mind, and hoping that most readers will, on reflection, be convinced of its validity, let us attempt to situate the new theory in a wider biological context. In order to do so, it may be helpful to look briefly at certain distinguishing characteristics of various other terrestrial organisms.

Now it has been perceptively remarked that new technology always seems like magic to a culture that has not yet acquired it. We did not suspect until this century (and could not have done because the relevant technology was far beyond our capacity even to imagine) that bats and dolphins use the highly sophisticated technique of echolocation in order to hunt and to move about safely. If our pre-scientific ancestors had been told that these creatures continuously bounce sound waves off objects in order to navigate, they would have initially considered this to be impossible and, if convinced that it really took place, then certainly 'unnatural'. Any creature that could function in such an extraordinary way would have seemed to them sinister and even diabolical. At best they would have found it hard to think of such animals as belonging in the same creation as, say, 'little woolly lambs'.

In fact, the nocturnal activity and the jerky flight of the bat really have generated superstitious horror throughout history. It is only in the present century that we have been able to perceive that the technologically advanced, and very efficient, methods of hunting and navigating which bats employ are intrinsically no more alarming than any other biological attribute. The motion of bats, we now recognize, is just as 'natural' as the movement of a colt trotting after a mare or of an otter swimming in a stream.

Modern science has shown that evolution usually finds the most efficient technology that can be applied to particular organisms in order to maximize their chances of survival. Nature's technology has, of course, until the last century or so, been consistently ahead of our own and usually ahead of our powers even to comprehend. It is only since the last World War that we have learned to harness this mysterious technology of echo-location for our own purposes.

Long before human beings had learned to wield clubs and spears, certain fish and reptiles were capable of using electricity for both hunting and self-defence. Fish beat us to this useful technology by a hundred million years. Migrating birds, as well as certain fish and marine mammals, can navigate using a great variety of sophisticated techniques. These include in-built magnetic compasses located in special cells, precise biological clocks, stellar and solar orientation devices and perhaps a kind of neural map.

We could, of course, instance many more examples of biological inventiveness anticipating human discovery. But enough has doubtless been said to suggest that the use of 'sophisticated technology' by smaller-brained fellow beings has not compelled us to modify our views of the world. Discovering that bats use echo-location or that eels can electrically stun a swimming man has not burdened our minds with a doom-laden sense that nature is no more than an assembly of biological robots.

We should not, therefore, succumb to culture shock on discovering that the technique which evolution has devised for supplying the vast information processing needs of the gigantic human brain is more relentlessly automatic than we had previously suspected. We may also find it helpful in coming to terms with this new and radical theory if we recall that 'little woolly lambs' have minds which, however rudimentary, are structured in precisely the same way as our own. They process data in sleep to provide a store of useful guidance during waking periods.

What all non-human species lack, and what gives our own mind powers far in excess of any others, is of course language. It is the use of language for immensely improving the brain's capacity for information storage and retrieval, and crucially for enabling raw data to cohere into self-consciousness, which categorically distinguishes us from all other living things.

We will now consider another aspect of the significance of the Twin-Data-Stream Theory and one which should provide a buttress for human self-esteem. Almost all thinkers — and indeed almost all human beings who think at all deeply about our nature — experience a need to believe that each one of us is unique. Except for extreme cults or systems which regard individuals as of inferior value to groups, religious and political systems are always based on the premise that every living person is critically different from all others and thus has a personal identity (sometimes called a 'soul') which is of potentially inexaggerable value.

In a sense this concept of 'unique value' is the *sine qua non* for the establishment of any kind of decent social order since without it human beings may be considered to be both interchangeable and expendable. The twentieth century has seen a number of regimes which have treated individuals as mere adjuncts to ideologies. Only the doctrine of the incalculable and unique value of every individual can durably provide the basis for decent administration and social justice.

Nevertheless, it has not always been easy to justify the assertion of human uniqueness. After all, is each ant in an ant hill unique? Does each seed from a blown dandelion manifest special and unparalleled qualities? Conceptually it can indeed be shown that each cell in every organism is an individual biological entity. But basing a plea for the irreplaceable value of human beings on a kind of uniqueness that can only be demonstrated in a laboratory does not provide a very telling political or philosophical argument.

In terms of the Twin-Data-Stream Theory, however, it is possible to maintain categorically that every human being really is functionally unique. This is because each human mind is the product of a 'one off' set of psycho-historical circumstances. As a corollary of this fact, it can be asserted that the value of each individual, as well as the contribution to society that he or she may make, must be inherently incalculable and potentially invaluable. It therefore becomes logically impossible to maintain that any group or any individual is expendable in the service of implementing an ideology.

This proposition may seem to generate the paradoxical notion that extremist and Fascist cultures, which may deny common humanity to whole categories of human beings, must themselves be desirable. The truth is that they, or at least the individuals that compose them, really are, at least potentially, valuable. This is because the vast resources provided by each individual archival memory enshrines the permanent possibility of radical change. The reformed sinner, the destroyer turned creator, the fascist become humanitarian are all common figures and issue from the immense variety and riches of the human archival memory.

Let us now look briefly at some of the more general implications of the Twin-Data-Stream Theory for human civilization. I am, of course, aware that my theory represents speculative science rather than laboratory science but maintain that this consideration is of small importance as compared to the question of whether or not it is true. It seems to me that if laboratory and other controlled experiments ultimately prove the theory to be substantially accurate then it is bound ultimately to have implications, and sometimes profound ones, for almost every aspect of human culture.

To take one branch of cultural activity almost at random, let us briefly look at the theory's implications for education. These implications strongly suggest, against the thrust of current educational theory and practice, that by far the most effective form of education is the long-despised technique of 'learning by rote'. This is because the labour of 'memorizing' facts ensures that the facts become a permanent constituent of the mind and are thoroughly integrated into the wider consciousness.

Multiplication tables, for example, if known by heart will endow the entire mind with an additional data processing dimension that can never be duplicated by consulting calculators or looking up tables. By memorizing poetry, a human being does not so much get to 'know' the works of a particular poet but in due degree actually becomes that poet. The memorizer's own mind will henceforth be fused with the mind of the poet and the latter will provide continual illumination, and enhanced powers of verbalization, throughout life.

The Twin-Data-Stream Theory thus endorses the validity of the classical, pre-modern view of the mind as being, at birth, virtually a '*tabula rasa*'. The human mind needs to be stocked with knowledge in order to become 'knowledgeable'. However, the new theory differs from the conventional '*tabula rasa*' belief in that it states that the mind is not only capable of being stocked but that the knowledge with which it is stocked will actually become a mental constituent of the individual.

The modern dogma of 'bringing out' innate creativity is exposed as a fallacy. Until knowledge has been first incorporated into the human mind there is simply nothing to bring out. Such concepts as 'the natural creativity of the child' are shown as pious dogma without any real meaning, a refutation which might come as a relief to many teachers who, after years of trying to coax green shoots from unseeded ground, may welcome a rethink of this fundamental aspect of educational theory.

If the theory has implications for education then it also has implications for sociology. And if for sociology then certainly also for the arts, and so on across the whole range of human mental activity. Clearly it would be not only over-ambitious but ludicrously premature to attempt here to make even a start at charting paths and currents of influence. I am naturally convinced that my Twin-Data-Stream Theory is broadly correct but equally convinced that it is certain to undergo some, and perhaps a large, degree of modification as a result of fine-tuning derived from laboratory testing. It will be time enough, if the broad central propositions of the theory sustain skeptical examination, to begin seeking implications for this or that branch of learning.

We must now make a very tentative start towards relating the Twin-Data-Stream Theory to the age-old problem of free will and determinism. Which of these classic alternatives is the most likely, in the context of the new theory, to express our fundamental orientation in the universe? At first glance it might seem that the Twin-Data-Stream Theory tilts the balance of probability in favour of determinism. After all, the theory states that our minds are products of the confluence of two data streams. It also asserts that both of these streams are generated by mechanical or quasi-mechanical external circumstances. It would seem a clear corollary of this analysis that our mental lives must, at least to a substantial degree, be controlled by external forces.

It remains a fact, however, that we do not feel as if

either our thoughts or our actions are rigidily determined. In other words, we remain psychologically locked into the traditional paradox of experiencing both a sense of freedom and an intellectual conviction that freedom is impossible. Is there any way to reconcile these diametrically opposed 'certainties'?

Let us first ask: is the psycho-historical situation of human beings significantly different under the new theory than it has been under older theories of how the human mind works? It is my belief that it is not. In the case of all theories of the mind, the same laws apply to biological organisms. No matter how free their mental processes may subjectively seem to them to be, it is impossible for such organisms to deploy any true freedom of will, and hence of action, in a universe that is structured in the form of a space-time continuum.

In other words the laws of cause and effect that govern the behaviour of matter in such a universe are not abrogated in the case of biological organisms simply because they are biological organisms. Put more simply, freedom of action is no more possible for living matter than for inanimate matter. The performance of any mental or physical act by a living organism is constrained by the same nexus of causality that governs insentient physical motion such as the interaction of fundamental particles and the orbital movement of planets. Even seemingly irrational or perverse behaviour — even behaviour that deliberately attempts to evade causal choices — is still bound into the same inflexible web of causality.

There is, it is true, an objection to this blanket assertion which readers equipped with a sophisticated understanding of modern physics might make. This is the fact that, in quantum physics, true indeterminacy does seem to be possible in certain circumstances. Whatever the truth about this, however, it is irrelevant to the question of whether the new theory tilts the balance of probability in favour of either free will or determinism.

It is also the case that free will, even if it were a characteristic of the universe as a whole, would be virtually impossible to isolate at the level of the individual human being. It would indeed be no more detectable at our mental level than it would be at the level of a blood corpuscle inside a human body. And yet, if free will really were a function of the whole universe, then both man and corpuscle would be able to deploy some trace of it, no matter how minute.

But this concept does not offer human beings much of an escape from the toils of determinism. The rather dispiriting truth must be stated that if men and corpuscles possess some pinch of free will then so do stones and atoms. If free will is an attribute of the universe to any degree at all then it must permeate the entire space-time continuum. There could be no 'no go areas' for free will. But clearly free will which operates at a level shared with inanimate matter confers upon human beings only a negligible degree of freedom.

But now, having apparently slammed the door on any hope of human beings having free will we can perhaps open it again. For the truth is that from the point of view with which we actually experience the universe, as distinct from the point of view with which we describe and analyze it, we really do have free will. What is more, this quality of free will, which actually governs all our choices, cannot rationally be denied to us by any argument available to beings that, like us, inhabit sequential time.

The key to the contradiction resides in the fact that human beings, and any other species capable of surviving in a universe structured dimensionally as our universe is structured, are constrained to live their entire lives on the borderline between the domains of determinism and free will. We call this frontier 'the present moment', and the two regions that are separated by this rolling instant are those of the past and the future. The past for us, as for any other any intelligent organisms inhabiting a universe having the same dimensional characteristics as our universe, is the realm of determinism while the future is that of free will.

It is relatively easy for any thoughtful individual to survey retrospectively his own history, or mankind's history or indeed the universe's history and detect enough evidence of the causal factors underlying them to be convinced that the universe is deterministic. It is not possible to apply the same procedure to the future. Ahead of the advancing 'present moment' is a region in which, for mind, causality has, other than in purely mechanistic spheres, only a negligible predictive power. We thus see that, in our universe, we are a kind of time-hybrid that perpetually links a deterministic past with a non-deterministic future.

So although it may indeed be true that what human beings have always understood by the phrase 'free will' is in fact the illusion of free will this perception should not generate any philosophical *angst*. It was just as true under previous theories of the mind as it is under the new one. But much more importantly, the basic fact is that if we seem to ourselves to possess free will then, for all practical and rational purposes, we really do possess it. Seeming to have free will is all that can ever logically have been meant by the statement that we possess it. Only an observer outside our universe — that is beyond the space-time continuum — would be in a position to deny that we have free will — and such an observer is impossible.

Similar common sense considerations can be evoked to refute some of the more bizarre science fiction hypotheses mentioned earlier in this chapter. Human beings cannot, for example, 'really' be computer-generated entities existing in a virtual-reality environment because the proposition cannot, in terms of experienced reality, ever be confirmed. If it were demonstrated that we are mere projections then, if the demonstration were convincing, our beings and history would instantly dissolve, and, if unconvincing, the demonstration could not crucially affect the integrity or nature of our perceived universe.

Thus we can be certain that we are inhabitants of a material planet precisely because we seem to ourselves to be inhabitants of a material planet. Calling ourselves 'computer-generated entities' cannot affect our experience of reality. It can therefore only ever be a theory without content since the content, should it ever be supplied, would immediately annihilate the originators of the theory. There are, therefore, no valid grounds for lamenting any supposed limitations to our alleged freedom of will that may seem to be implied by the theory of mental structure outlined in this book. We remain just as free — or, from a different perspective, unfree — under the new dispensation as we were under the old.

However, it is valid to make the encouraging observation that the Twin-Data-Stream Theory binds us much more intimately into the physical processes of our environment than any previous theory of the mind has done. This apparent constraint paradoxically provides us with increased opportunity to interact purposefully with that environment. It means that we are necessarily executives operating and reasoning in the inescapable context of experiential reality. This in turn means that we are true participants in the cosmic process and not just untethered intelligences lacking any strong, organic connection with the rest of the universe.

I will conclude this short book with the assurance that, after some initial distress about the implications of the Twin-Data-Stream Theory, I have come to believe that it offers hope in many fields of human thought and activity, perhaps most immediately in the diagnosis and treatment of pathological mental states. It is also my belief that this hope far outweighs any threat that the theory might initially seem to pose for human self-esteem.

I am not a professional scientist but I share with most scientists the firm belief that the truth is invariably good.

It is, of course, a fact that scientific discoveries do not always reveal their positive character when they are first made public. Indeed, some of them really may pose challenges to our physical survival and others cause alarm and even social disorder while their wider implications are being assessed and assimilated.

In the long run, however, truth, if its implications are fully comprehended by society and sensibly incorporated into its administration, will always increase our ability to interact fruitfully with the universe. I feel genuinely confident, therefore, that should the Twin-Data-Stream Theory emerge from rigorous laboratory testing as substantially accepted truth it will prove to be of benefit to mankind.

# Brief Bibliography


Asimov, Isaac, *Asimov's Guide to Science*, Penguin Books, 1987.

Clare, John, 'Written in Northampton County Asylum', *Oxford Anthology of English Poetry*, Oxford University Press, 1990.

Davies, Paul, *Are We Alone?*, Penguin Books, 1995.

Dennett, Daniel C., *Consciousness Explained*, Allen Lane, 1991.

Empson, Jacob, *Sleep And Dreaming*, Harvester Wheatsheaf, 1993.

Freud, Sigmund, *Standard Edition*, Hogarth Press, 1950.

Hofstadter, Douglas R., *Godel, Esher, Bach: An Eternal Golden Braid*, Penguin Books, 1986.

Joyce, James, *Ulysses*, Penguin, 1968.

Keller, Helen, *The World I Live In*, Century Co., New York, 1908.

Koestler, Arthur, *The Sleepwalkers*, 1959.

Magee, Bryan, *Confessions of a Philosopher*, Weidenfeld & Nicholson, 1997.

Masefield, John, 'Sea Fever', *Collected Works*, Heinemann, 1923.

Pascal, Blaise, *Pensées*, Harvill, 1962.

Penrose, Roger, *Shadows of the Mind*, Oxford University Press, 1995.

Sacks, Oliver, *The Man Who Mistook His Wife For A Hat*, Picador, 1985.

Searle, John, *Minds, Brains, and Science*, BBC Publications, 1984.

Shakespeare, William, *Henry IV, Part II*, Signet Classic, 1965.

Sidney, Sir Philip, 'Astrophel & Stella', *Sonnets*, Clarendon, 1971.

Tennyson, Alfred, Lord, 'In Memoriam XXXVI', *Collected Works*, Dent, 1974.

Tolstoy, Leo, *Anna Karenina*, Penguin, 1954.

# Index


Alpha waves, 32, 75, 80
Amnesia, 92, 94, 139–43
  infantile, 105–6, 113
Ancestors, proto-human, 22–3, 87–8
Animals, 35, 37, 83–6, 96–111 *passim*
  brain, 85, 96, 98, 99, 102
  mind, 83–5, 95, 97, 109, 110, 113, 149
Arts, 99, 152
Asimov, Isaac, 14, 15
Autism, 132–3

Babies, 84, 86, 101, 105–6, 113
Bats, 147, 148
Brain, human, 10, 11, 15–29, 35–50, 73–5, 78–83 *passim*, 86–7, 105, 113, 115, 121, 149
  functions, diurnal, 19, 35–6, 64–5, 78
  'drone', 78, 86
  nocturnal, 17–18, 36–41 *see also* Sensory data

'Calculating wonder' *see* idiots savants
Chimpanzees, 84, 88
Clare, John, 130–1
Constellation, 95
Co-operation, 87–8
Creativity, 152
Culture/civilization, 47, 55, 85, 99, 147, 151
'Cyborg', 77

Davies, Paul, 61
Day-dreaming, 29
Death, 99, 104
  brain, 75
Deity, 108
Dennett, Daniel, 14, 82
Determinism, 110, 152–5
Dogs, 35, 37, 49, 120, 126, 137
Dolphins, 85, 98–9, 144, 147
Dozing, 45
Dreams, 9–18, 36, 38–40, 42–8, 50–64, 69–72, 91, 93, 125, 139
  content/nature, 13–15, 19, 39, 40, 42–3, 45, 46, 52, 57–61
  distortions in, 44
  formation, 58–9, 61, 64, 69, 71–3
  functions, 53, 54
  incorporation, 53–4
  interpretation, 12, 46, 51–3, 55, 58, 61–4
  non-memorability, 69–70, 72
  'waking', 47, 128, 129
Dualism, 110–13 *passim*

Echolocation, 147–8
Education, 151–2
Electroencephalography, 13, 32
Electro-oculogram, 48
Empson, Jacob, 14–15, 48, 53, 125–6
Error, 24, 25
Evolution, 55, 86–7, 113, 120, 144, 148, 149
Experience, 98, 119, 137
  learning from, 98

Feynman, Richard, 42
Flicker epilepsy, 83
Free will, 110, 152–6
Freud, Sigmund, 53, 54
Friendship, 99, 102, 104

'Ghost in the machine', 74–5, 104, 105
Gleick, James, 42

Hallucinations, 39, 127–30 *passim*
Hofstadter, Douglas, 14
Hunting techniques, 88, 148
Hysteria, 138

'idea-chain', 29
'Identity check', 141–3
Idiots savants, 134–6, 138
I/not I distinction, 85, 99
Instinct, 99, 100
Intelligence, 85, 87, 121
Intention, 109, 111
'Interior monologue', 119
Interpretation, 97–8 *see also* Dreams

Joyce, James, 119

Keller, Helen, 99–100
Knowledge, 102–6 *passim*, 120, 151–2
Koestler, Arthur, 55

Korsakov's syndrome, 92–4, 131–2, 139–40

Laboratories, dream, 15, 43, 53
sleep, 48
Language, 17, 83–8, 94–104, 109–13, 116–20 *passim*, 149
'Liar paradox', 61
Love, 99, 102, 147

Magee, Brian, 102, 103
Malfunctions, 12–13, 24, 78, 125–44, 156
Manic depression, 138
Masefield, John, 145
Mathematics, 99, 102, 135
Memo signals, 78, 85, 141
Memory, archival, 23, 25–41 *passim*, 47, 54–9, 62–72 *passim*, 76, 79–83, 92, 94, 111, 114–17, 127–38 *passim*, 141, 143–4
short-term, 36–41 *passim*, 47, 49, 56, 75, 92, 128, 139
Migraine, 138
Mind, human, 17, 24, 27, 45, 73–86, 90, 94–7, 109–15 *passim*, 119, 151–3
'architecture', 30–4
generation, 10, 32, 33, 36, 70, 73, 74, 78–84, 113, 115–17, 133
subconscious, 20, 22, 24, 25, 81
switching off, 73–4, 76–7, 126
Monitoring, self-, 25–6, 30
sensory, 28–9, 31, 33, 78, 80, 81, 83, 93, 111

Narrative, 58, 97, 119 *see also* Story
Navigation, 84, 147, 148
Nerves/nervous system, 28, 76, 79–81, 105, 109, 112, 114, 115

'Out-of-body', 9–11, 19–20, 138–9

Pain, 104, 147
Parasitism, 104–5
Pascal, Blaise, 90
Penrose, Roger, 14
Personality, 17, 74, 77, 80, 140, 142
'Phantom limb', 80
Prophecy, 46, 51–2
Psychoanalysts/psychotherapists, 12–13, 52

Reflex action, 82, 96, 98, 120
Rest, 13–14, 55, 76

Sacks, Oliver, 92–4, 134–5
Schizophrenia, 93–4, 127–32
cataleptic, 131
Science, 13–15 *passim*, 55, 99, 104
Screening, 27, 32, 33, 36, 38, 66–8, 70, 79
malfunctions, 127–33
Searle, John, 115
Self-consciousness, 17, 84, 98–100, 107, 109, 113, 118, 120, 133
Sensory data, 28–32, 35–8, 41–5, 47, 54, 64, 82, 92, 96, 97, 113, 115, 128–31, 133
cross-referencing/filing/indexing/processing, 21, 31, 36–40, 42, 44, 45, 47, 49, 54–6, 59, 62, 64, 69–77 *passim*, 99, 114, 126, 128, 135–8 *passim*, 146, 149
relevance, 35, 37, 92, 116
retrieval, 21, 36, 38, 42, 127, 149
selection, 26–8, 35, 40, 66–8, 135
storage, 23, 32, 38, 41, 47, 84, 106, 113, 135, 149
Shakespeare, William, 118, 119, 145
Sleep, 12–18, 36–8, 44, 46–64, 73–8, 145–7 *passim*
deprivation, 47, 128
protection, 54, 55
REM, 13, 47–50, 73–5, 84, 126
Sleepwalking, 49, 125–6
Sociology, 152
Soul, 104, 149
Space, 85, 90, 91, 104
Story, 45, 90–1, 94–5, 121
'of me', 86, 88, 92–5, 97, 106, 110, 118, 119, 121, 132
Sydney, Sir Philip, 145

'Talking to ourselves', 85, 86
Thinking machines, 143–4, 155–6
Thoughts, 30, 65–7, 81, 100, 102, 103
'from nowhere', 19–26, 28, 34, 65
relevance, 25–7, 33
'rogue', 23–6, 33
Time, 43, 45, 82, 83, 104, 120
space distinction, 85
Tolstoy, Leo, 119
Topography, 21–3
Twilight period, 44–5, 54, 69
Twin-data-stream, 30–6, 47, 53–9 *passim*, 77, 83, 92, 97, 113, 115–18, 125–8, 131, 132, 138, 143, 145–52 *passim*, 156, 157
turbulence of, 30, 31, 83, 115
'Two I's', 9, 11, 39, 40, 53, 54, 139

Uniqueness, 149–50

Voices, 128, 130

Wordsworth, William, 145
Writers, 118–19